Sunset
GARDEN & PATIO
Building Book

By the Editors of Sunset Books and Sunset Magazine

Several outdoor building projects meet in this garden of earthly delights. Brick rectangles border a tiled ornamental pool; beyond it, pathway decks step across to a large garden shelter. Landscape architects: Kawasaki/Theilacker.

Lane Publishing Co.• Menlo Park, California

Staff Editors:
Don Rutherford
Susan E. Schlangen
Scott Fitzgerrell

Special Consultants:
William Louis Kapranos
Landscape Architect

Peter O. Whiteley
Associate Editor,
Sunset Magazine

Photo Editor:
JoAnn Masaoka

Design:
Roger Flanagan

Illustrations:
Bill Oetinger

Handsome fence and gate of prestained redwood enclose a redwood deck built over an existing concrete porch. Matching trellis shades end of deck that serves as path to house entry. Landscape architect: Donald G. Boos.

Photographers

Jack McDowell: 2, 6, 7 top, 9 top, 11 top right, 12 bottom, 13 top, 15 top, 17, 19 top, 20, 21 bottom, 24 top right, 25, 26 left, 27 right, 31 top, 32 top center left, 32 bottom far left, 32 bottom far right. **Stephen Marley:** 1, 3, 5, 8 top, 10, 11 bottom, 16 top, 19 bottom, 21 top, 22 bottom, 23, 24, left, 24 bottom, 26 right, 27 left, 28 bottom, 29 top, 30 left, 32 bottom center left. **Ells Marugg:** 18 top. **Norman A. Plate:** 16 bottom. **Bill Ross:** 32 top center right. **Rob Super:** 9 bottom, 12 left, 15 bottom, 32 bottom center right. **Tom Wyatt:** 4, 7 bottom, 8 bottom, 11 top left, 13 bottom, 14, 18 left, 22 top, 28 left, 29 bottom, 30 right, 31 bottom, 32 top far left, 32 top far right.

Design credits for photos on page 32
Bruce Andrews: bottom far right. Thomas L. Berger Associates: top center right. Fisher-Friedman Associates: top far right. William Louis Kapranos: bottom far left. Singer & Hodges: top far left. Jeff Stone: bottom center left. Taro Yamagami: top center left.

Cover: White-painted shade structure links house and patio in a graceful design that lends a touch of tradition to modern outdoor living. Straightforward design of both shade structure and paving made them easy to build. Landscape architects: Armstrong & Scharfman. Photographer: Jerry Fredrick.

Editor, Sunset Books: Elizabeth L. Hogan

Sixth printing October 1988

Contents

Deck with built-in bench, planters, and light shows skillful design and careful attention to detail. Design: Bob Waterman.

Used-brick walls and steps combine with exposed-aggregate concrete paving in this masterful masonry composition. Landscape architect: Thomas E. Baak.

Special Features

DESIGN IDEAS

Colorful inspiration for pavings, walls, fences & gates, decks & overheads

Life in the great outdoors—or even in a great back yard—is one of those rare pleasures enjoyed equally by people of all ages. Hardly anyone can resist at least thinking about outdoor building projects—from children building their daydreams in the dirt to adults doing basically the same thing on a different scale.

Whether you contemplate a sweeping orchestration of pavings, walls, fences, and decks, or would just like to do a little puttering, this book is for you. In it you'll find all sorts of ideas and projects—from the modest to the magnificent.

First, to get your imagination going, we feature a gallery of ideas in color—examples of what real people have done in real situations.

The gallery has four parts: paving (pages 6–13), walls (pages 14–19), fences and gates (pages 20–23), and decks and overheads (pages 24–31).

As you look at the photos, you'll note that most show elements of an overall garden or patio project in addition to their featured subjects, making it possible for you to glean ideas from "the big picture," too.

Ideas and inspiration, or utterly practical information—whichever you're looking for, you'll find it here in abundance. This book was designed for doers *and* dreamers; read on and see.

Classic style

Large-scale poolside project contains many adaptable ideas. Stone planters have comfortable wooden tops for seating; style of heavy trellis is echoed in simple deck railing beyond. Cedar deck is neatly butted into mortared stone paving at point where hillside begins to slope away beneath. Landscape architects: Singer and Hodges.

Secret garden

Brick basket-weave path, board fence with lattice top and integral trellis, and gate with grille of decorative turnings are major elements of this storybook garden entry. As twilight falls, custom-built lights show the way. Design: Bob Waterman.

Paving:
For carefree outdoor living

Creative combination

Exposed aggregate concrete attractively frames panels of mortared brick paving. Raised area serves as a path between gate and house; flanking it are small patios for sitting and gardening. Landscape architects: Lang & Wood.

Modular design

Redwood-bordered concrete squares feature exposed aggregate surface for beauty and durability. They're part of a modular design that also incorporates a concrete slump-block firepit and bordering wall. During construction, the redwood grid divided work into stages; with patio complete, it becomes an attractive design element. Design: MLA/Architects.

Simplicity itself

Pea gravel covers bare earth in this intimate garden seating area. Border of 2 by 4s (set vertically a foot or so into the ground) meanders along its edge, creating a low planting bed and confining gravel. Landscape architect: Kenneth Pederson.

Paving:
Long-wearing patio floors

Earthy elegance

Terra cotta tile gives this dining-room patio a sophisticated informality. Tiles were laid over a concrete slab; matching tiles, in a larger size, cover the adjoining concrete seating wall. Architect: John E. MacAllister.

Brick basics

Brick patio in basket-weave pattern was laid using the dry-mortar method: Bricks were placed on a sand-cement bed, then dry mortar was tamped between them. Fine sprinklings with water set the mortar to complete the job. Owner used a masonry saw to cut bricks to fit around the supports of the finely crafted tub, table, and bench. Design: Bob Waterman.

Surface treatment

Irregular slabs of tawny sandstone, mortared over an existing concrete slab, floor this patio and spa deck. Complementary planters of the same stone flank entry to spa. Design: Bob Rubel & Associates.

Building blocks

Concrete blocks can appear to striking advantage in patio settings. Here, 2-inch-thick blocks were laid on a sand bed; they're secured by redwood timbers bolted to small concrete footings.

Paving:
Steps in the right direction

Easy climbing

Slabs of exposed aggregate concrete ascend gently to a tree-shaded deck. Sectional construction like this (each slab was cast in its own frame of 2 by 6s) is ideal for weekend-by-weekend builders. Landscape architect: Thomas E. Baak.

Working on the railroad

Steps made of paired railroad tie sections link two decks set at different levels. Virtually indestructible, ties are easy to work with. Their dimensions make it simple to build steps with ideal riser-to-tread ratios. Landscape architect: Michael Painter.

Piecework

These sturdy steps were laid stone by stone in mortar beds placed directly on soil; no slab foundation was required. Following a stable hillside's natural contour, steps were built on firm, cut (not filled) soil. Landscape architect: William Louis Kapranos.

Embellishment

Brick trim applied to a set of functional concrete steps makes a decorative edge. Pitted concrete surface was formed by rock salt embedded in the fresh concrete, then washed out once the surface hardened. Landscape architect: Thomas E. Baak.

Paving:
Traffic patterns

Down to earth

Winding path of adobe blocks was laid in basket-weave pattern with butted joints. These earthen blocks, easily cut with a handsaw, rest on a bed of stone screenings.

Progress in stages

Smooth stones separate slabs of salt-finish concrete in this easy-to-build garden pathway. Movable forms were used to cast individual slabs, breaking up the work and making it easy to accommodate changes in grade. Landscape architect: Thomas E. Baak.

Nice to come home to

Beautiful entry walk shows how creative brickwork can be. Combination of herringbone paving with jack-on-jack edging, large planter, and steps (one of which extends into a low retaining wall) is elegant and harmonious. Landscape architect: John J. Greenwood.

Cobblestones in miniature

Smooth stones were pushed into wet concrete like raisins into dough to make this striking driveway surface, a variation on the exposed aggregate concrete technique (see page 38). Also notable are handsome flagstone divider strip and inconspicuous handling of the control joint seen in the foreground. A coat of sealer perks up stones' color. Landscape architects: Singer & Hodges.

DESIGN IDEAS **13**

Walls:
Marking the boundaries

Symphony

Graceful brick wall displays an unusually harmonious combination of distinct elements. Stretcher bricks form openwork screen atop common-bond base, admitting light and air. Screen is topped by decorative header, soldier, and stretcher courses. Two pilasters act as gateposts; others, not shown, reinforce the wall at intervals. Architects: Fisher-Friedman.

Artifact

Like a remnant of some vanished civilization, this stone garden wall projects an aura of simple dignity and beauty. Built of sedimentary stone, wall has thin mortar joints, raked to give appearance of dry-laid construction. Landscape architect: Kenneth W. Wood.

Warm embrace

Earthy adobe wall wraps a sheltering arm around a small patio. Casual mortar work is appropriate to this material—and easily achieved by an amateur. Mud blocks contain an asphalt stabilizer, and don't need the traditional coat of waterproof plaster.

Walls:
Making a place for plants

Dividing line

.Sinuous stone wall forms a sturdy barrier for planting bed above, a clear-cut border for paving below. Rugged lava rock makes an effective foil for a boisterous crowd of marigolds. Landscape architect: Thomas E. Baak.

Fortress for flowers

Interlocking wooden ramparts ascend a low hill in this garden, forming a series of planting terraces. A coat of semitransparent stain protects the timbers and gives the assembly a sleek, unified look. Design: Charles S. Randolph.

Instant atmosphere

Adobe block, evocative of the old Southwest, is always a good choice for garden projects with a Spanish accent. Large block size means a low wall like this one can be completed in fairly short order. These blocks are "stabilized" and need no further waterproofing. Architect: Donald G. Boos.

Touches of distinction

Simple planter of textured brick displays subtle design details. Double corner at right keeps planter from projecting too far out at house corner; top's rounded bull-nose bricks soften the overall effect and are less prone to chipping than square-edged bricks. Planter works nicely with sand-bedded paving of variegated, flashed brick. Landscape architects: The Peridian Group.

Walls:
Keeping the earth at bay

Roughhewn beauty

This dry-laid stone retaining wall relies on a slight backward tilt (or "batter") and its own weight for stability. Its height called for a building permit before work began. Crevice plantings soften the near-cyclopean look.

Natural occurrence

Fieldstone wall emerges from a group of strategically placed boulders. Mortared stones provide a solid barrier between lawn above and planting bed below. Landscape architects: Singer & Hodges.

Graceful impressions

The strength of a poured-concrete retaining wall is often purchased at the expense of a drab surface. This wall is enlivened by impressions left by rough form boards that molded the wet concrete. About three feet high, it's a feasible project for a careful do-it-yourselfer. Landscape architect: Paul McMullen.

In the folk tradition

Undulating brick retaining walls keep lawn and planting beds in place, making room for a set of gently rising steps. Folksy combination of used and flashed brick—and even an occasional stone—sets off informal curves. Landscape architect: Woodward Dike.

Fences & gates:
Beautiful barriers

Hanging garden

Giving new meaning to the term "garden fence,"
here a garden and fence are one. Tapestry of lush
plantings hides a metal frame supporting two layers
of chain-link fencing. Cavity between layers was
stuffed with moss and soil, then planted. Built-in drip
irrigation system handles watering and fertilizing
chores. Landscape architect: Bob Truskowski.

Integral look

Sleek clapboard fence encloses a front yard remodeled for privacy (see page 24 for another view). Fence sits atop a concrete foundation and is framed and sheathed much like a house wall. Garage was re-sided also; careful matching of clapboard pattern from fence to garage door contributes to "seamless" effect. Landscape architects: The Runa Group, Inc.

Light touch

White paint and a lacy top dress up this fence built primarily of 1 by 6 boards on a simple frame. Graceful trim includes bands of molding and panels of lattice. Matching gazebo repeats elements of fence design. Design: Richard Murray.

Fences & gates:
Filtering light, air & view

Palisade on poles

Peeler poles provide support and visual punctuation for arched panels of redwood 2 by 2s. Designed as a front-yard privacy screen, fence admits plenty of light and air, yet provides a sense of enclosure. Landscape architect: William Louis Kapranos.

Tradition

Crisp latticework fence and gate form neat line of demarcation between driveway and back yard. Minimal barrier is more symbolic than functional; its own traditional charm is its reason for being. Landscape architect: Thomas E. Baak.

Private retreat

Tucked away in one corner of an entry deck, cozy
cul-de-sac is shielded from street by simple reed
fence. Sturdy frame of 4 by 4s supports panels of
reed screening, which can be easily replaced when
they become worn or discolored.

Entry drama

Elegant articulation of elements
marks this strong-lined yet unfor-
bidding garden entry (photo
shows its inner side). L-shaped
deck and sheltering shade structure give an almost ceremonial
sense of entrance, while openwork gate and flanking "wing-
wall" fences entice viewers with
glimpses of lush, serene garden
within. Landscape architects:
Kawasaki/Theilacker.

Decks & overheads:
All around the house

(see page 21)

Reforming a narrow side yard

Narrow lots often mean even narrower side yards. But thanks to a new split-level deck, the space shown here has gone from hard-to-use to hard-working. Changes in level and angle help break up the awkward space, creating several focal points along the deck. Landscape architects: Todd Fry & Associates.

Privacy up front

This front-yard deck is the key to an effective face lift that turned a small, sunny, but too-public space into an attractive place for outdoor living and entertaining. A new fence (see page 21) encloses the remodeled yard. Landscape architects: The Runa Group, Inc.

Getting there

Decks can double as pathways. Here, a staggered series of platforms passes through lush, undisturbed ground cover. Decks keep foot traffic—and plants—within bounds, while enhancing overall garden design. Low-voltage lights make unobtrusive nighttime guides to a change in level between two platforms. Landscape architects: Kawasaki/Theilacker.

A special place out back

A deck can sometimes solve the problem of what to do with the last bit of back yard that often goes unused. Built around a large tree, this deck, with its integral benches and planter, transformed a forgotten corner. Landscape architect: Jim Coleman.

Decks & overheads:
Made for shade

Sheltered walkway

Gracious entry walk is first shaded by a simple trellis, then fully sheltered by a shingle roof. Trumpet vines stitching their way through trellis will soon provide a canopy of flowers. Landscape architects: Todd Fry & Associates.

Careful engineering

Massive trellis rests solidly on rugged uprights, each made of four interlocked timbers. With so much weight aloft, rigid connections are a must to prevent sway—hence the sturdy bolted (not screwed or nailed) construction. A coat of semitransparent stain protects wood without obscuring its attractive rough-sawn texture. Landscape architect: Kenneth W. Wood.

Craftsman's touch

Elegantly simple shade structure protects an enclosed patio reached through gate in foreground. Notched joinery of the structure allowed assembly with a minimum of metal fasteners, for a clean, well-crafted look. Architect: John E. MacAllister.

Filtering the sun

Lattice panels run the length of this house, forming a shaded gallery for walking and sitting. Because structural stability comes from the house itself, construction was simple despite the project's large scale. Color scheme matches the house. Landscape architects: The Peridian Group.

Decks & overheads:
Decks for pools, tubs & spas

Attractive surroundings

Deck meets pool edge perfectly for a sleek, flush-mounted look. Wraparound bench is functional in more ways than one: sections of its top hinge open for access to storage within. Fiberglass pool was installed by a crane that simply lifted it over the house. Landscape architect: Thomas E. Baak.

Crafty coopering

Small decks surround a classic, barrel-like hot tub. A taper jig and table saw made short work of tapering deck boards, speeding construction of this minimum-space design. Design: Roger D. Fiske.

Cover-up

A worn, outdated pool deck was a source of frustration for its owners—until it became a source of inspiration. Replacing it would have been arduous and expensive, so they simply covered it up. Result: a neat wooden deck-and-step combination enhancing both pool and yard. Architects: CZL Associates.

Geometry lesson

Twelve-sided spa forms the nucleus of a tri-level radial deck designed to handle large parties and solitary soaks with equal finesse. Upper level, a "public" space with continuous integral bench, extends inward to become a seat near the spa. Latticework railings and bench surround the spa, creating a more intimate space. Design: Dan Klitsner and Kirk Von Schwindeman.

Decks & overheads:
Outdoor rooms

Collaboration

Airy pavilion anchors one corner of a team-effort garden design. Ambitious owners built pavilion, deck, wall, and koi pond, but relied on a landscape architect's plans to guide them. This pooling of resources resulted in a professional design meticulously executed—at considerable savings. Landscape architect: Woodward Dike.

True romance

Gazebos hint at more shelter than they actually provide—romance, not practicality, is the key to their appeal. As with most gazebos, the real function of this one, with its lacy latticework arches and railings, is to offer a place of irresistible charm for quiet enjoyment of the garden. Octagonal structure is bolted to a brick and concrete base; roof is made of redwood bender board and cedar shingles. Design: Gazebo Nostalgia.

Poolside pavilion

A rectilinear timber frame supports the lattice panels of this shady poolside retreat. Long spans overhead provide a large open area below. Triangulating frame members work both as decorative accents and as all-important structural braces. Landscape architect: John J. Greenwood.

Command post

Handsome work center dominates one end of garden. Structure consists of two large counters beneath a shading roof made of 2 by 2s laid over 2 by 10 rafters with decoratively cut ends. Roof is braced by steel rods and turnbuckles running diagonally from corner to corner. Near counter has raised cap to hide work surface from view and provide display space for potted plants. Landscape architect: Emery Rogers.

PAVING

Eight surfaces for patios, walks, and steps

Concrete (see page 36)

Exposed aggregate concrete (see page 38)

Concrete blocks (see page 39)

Brick (see page 40)

Flagstone (see page 44)

Tile (see page 45)

Adobe (see page 46)

Quick & easy (see page 47)

Whether it's a brick or flagstone patio for entertaining, or a concrete walk for hauling out the trash on collection days, paving is one way to greatly enhance the beauty and value of your property. Discussed on the following pages are seven of the most popular surfacing materials, along with specific construction pointers.

Before you choose a paving material, be aware that each material has strong and weak features. Here are some points that you'll want to take into consideration.

• Surface texture affects comfort underfoot, ease of maintenance, and safety. Smooth pavings, for example, are easy to maintain and are good for entertaining areas. But when wet, they may be slippery.

• Appearance reflects your taste in color, texture, and pattern. Choose what you like and what best suits your life style—you're likely to live with it for a long time.

• Maintenance of paving surfaces involves hosing, sweeping, or blowing. Smooth surfaces are easier to maintain than rough ones; grouted joints are easier to clean than open ones, and they prevent weed growth.

• Durability is affected more by the climate than by the kind of paving material. In areas of extreme freezing, frost heave can move pavings, whether set in concrete or sand. Brick-in-sand paving is particularly prone to upheaval; you may have to rework the bricks every spring.

• Accessibility to the job site may affect your choice of materials. You probably won't use heavy paving blocks if you must carry them a considerable distance by hand.

Paving: Choices in materials

Once you've given some thought to basic considerations, you can judge each paving material on its particular merits.

Concrete. From a few simple ingredients—cement, sand, gravel, and water—you can make a hard, durable surface to suit any garden or patio. Concrete can be used to build anything from a small stepping-stone to a large slab—and in any shape you want (see page 36).

Depending on how you finish the surface, concrete can be smooth enough for shuffleboard or dancing, or rough enough to stop you from slipping on a wet, steep path. Concrete can also be seeded with attractive stones (see below). Careful planning and preparation are essential, though, and you'll need plenty of help if you pour concrete in a large area.

Exposed aggregate concrete. Smooth varicolored gravel makes this distinctive concrete finish one of the most popular choices for residential concrete paving (see page 38). In some cases, an exposed aggregate finish is made with cobblestones, river stones, or pebbles placed in wet concrete to create a cobbled surface.

Concrete paving blocks. You can buy cast-concrete pavers or make your own, casting them in wood forms or in place in the ground. Though concrete blocks are usually set in a bed of sand, they can be positioned directly on stable ground. (See page 39.)

Interlocking concrete blocks are a modular paving product. Available in various shapes for pleasing patterns, blocks are set in a sand bed and the joints are filled with sand.

Brick. Brick, available in many types and sizes, has become relatively expensive, especially when compared with concrete. But you can probably justify its expense when you consider its beauty, permanence, and maintenance-free character. You can set bricks in a bed of mortar, or simply in sand, in a variety of patterns (see page 40).

There are other advantages, too. The rough surface of many kinds of brick provides traction. Because of their size, you won't need any help to put bricks in place, and if necessary, you can easily cut them with simple tools.

Flagstone. This is one of the most durable outdoor surfaces available. Flagstone can be laid in sand or mortar or, if the stone is thick enough, directly on stable soil. The stones' size and weight add stability. Texture and color vary according to where the flagstone is quarried. For more on flagstone, see page 44.

Tile. For an extremely smooth, durable surface that's as hard as the toughest stone, try exterior tile (see page 45).

Tile resists abrasion and soiling, and is as easy to lay as brick (though it costs considerably more). When it's wet, tile—particularly glazed types—tends to be slippery unless it has a rough surface. You can set tile in sand, dry mortar, wet mortar, or wet mortar on a wood deck.

Adobe pavers. Stabilized with an asphalt emulsion, adobe pavers are waterproof and nearly as durable as brick. Offering a friendly, rustic appearance, these pavers are available in several different sizes and are set in sand with 1-inch joints filled with sand or dirt (see page 46). For those who live near a source for adobe, it is cheaper than brick; otherwise, the cost may be prohibitive.

Quick and easy pavings. If you're looking for an informal appearance with a minimum investment of time and effort, consider loose material (bark, gravel, decomposed granite, crushed rock or brick, redrock) or wood set on sand or gravel (see page 47). Inexpensive compared to hard pavings, these materials are easy to install, and they drain well.

Paving: Preparing a base

For most pavings, you must have a base of firm, well-drained soil and a setting bed of sand or gravel. The soil and setting bed provide stability for the paving. If you anticipate drainage problems, see pages 48–49.

Drainage

Any time you pave an area, you affect its drainage. Water will tend to run off even the most porous paving, such as brick in sand. Unless the area to be paved slopes naturally, you must grade it before paving, to prevent runoff from collecting where it will cause problems—against a house foundation, for example. You should provide a slope of about ¼ inch per foot.

If the soil on your paving site doesn't drain well, lay a 4 to 6-inch bed of 1-inch gravel to improve drainage. Be sure to allow for such a bed as you prepare your base. Try to avoid sending the runoff toward any place that is already boggy during a rain—to do so would only make the problem worse. Refer such a problem to a landscape contractor or architect.

Grading

For many homeowners, grading involves only minor changes in the contours of the ground.

Before you start any digging on a grading project, check with your utility companies to locate underground lines for gas, water, sewer, electric, telephone, and other services. If your home is not connected to a sewer, locate the septic tank and associated drainage field.

Though you may be able to grade your yard by guess and by eye, setting grading stakes and establishing a grid to guide you will make the task easier. To grade a large area like a front or back yard, first stake the perimeter off in intervals of 5 or 10 feet. (If the ground is irregular, use 5-foot intervals—smaller grids provide greater accuracy in grading.)

Determine your preferred level for the finished paving surface and mark it on one stake. (This should be at or above existing grade.) Attach a chalk line at this point and stretch it along a row of stakes. Level the line (using a regular level or line level designed to hang on the chalk line) and snap it to mark the stakes. Repeat for all stakes.

Adjust your marks to allow for a slope of at least 1 inch in 10 feet toward the direction you want water to drain (as recommended in "Drainage," preceding). Plan to have the low side of the paving end at or above the natural ground level, not below it. Using a chalk line, mark the stakes again. Finally, establish a grid by stretching mason's lines between opposite pairs of stakes.

To grade smaller areas such as walks, erect batterboards beyond each corner so that the boards' top edges are 6 inches above the desired paving height, and outline the area with a mason's line as shown below.

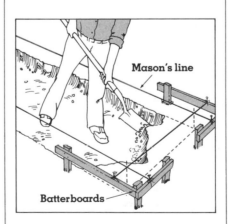

Once the grade is marked, excavate the area, allowing for the paving thickness plus the thickness of the

sand or gravel bed. If possible, avoid filling and tamping—tamped soil is never as firm as undisturbed soil, and it will probably settle, taking your paving with it. A square-ended shovel is useful for cutting through grassy areas and for keeping the excavation edges perpendicular. The excavation area should be about a foot wider than the planned size of the finished paving.

If the earth is soft and wet, tamp the graded soil (you can rent a commercial tamper). Don't worry if the depth is greater than planned or if you excavate too much; you can add enough sand or gravel to the setting bed to compensate.

Laying the setting bed

When you've completed the excavating, you're ready to lay the sand or gravel setting bed (if needed for the paving you plan to install). First, though, build forms or edgings (see page 35). These will act as a guide for the bladed screed used to level the sand or gravel.

Dump sand or gravel into the excavation and roughly level it with a shovel and a rake. Place the bladed screed on the forms or edgings and work it back and forth to level the sand or gravel. Add material to fill in depressions.

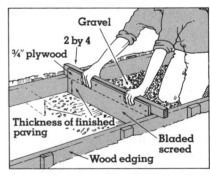

If you use sand, you'll need to dampen it and use a roller to compress it. Add more sand, rescreed, and repeat until the bed is the desired height.

Allow the sand or gravel to fill any gaps under the forms or edgings.

Paving: Edgings for patios, walkways & paths

Edgings hold the base and paving units of paved areas in place, make attractive divisions or grids within paved areas, and can neatly define lawns, flower beds, and other planting areas. When used to curb paving areas, edgings usually are installed after the base is graded (see facing page) and before the setting bed and paving are laid. The edgings' finished height should be flush with the finished paving's surface.

Brick-in-soil edgings

The quality of soil must be firm to hold easy-to-build brick-in-soil edgings securely.

In the drawing below, the soil has been cut away to show a row of "sailors" (bricks standing side by side). The full length of the bricks is buried to prevent tipping.

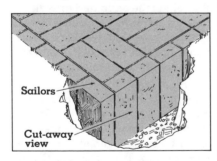

Sailors

Cut-away view

To install brick-in-soil edgings, first grade the area to be paved (see facing page). Dig a narrow trench around the perimeter, making it deep enough so that the tops of the sailors will be flush with the paving when it is in place. Then position the sailors, leveling their tops as you work. Pack soil against the outer perimeter of the edging bricks (as you set the paving bricks).

"Invisible" edgings

An invisible edging is actually a small, underground concrete footing that retains the paving units and base with no apparent support. Paving units are set into the concrete's surface to conceal the footing. This is a strong edging, particularly effective with brick-in-sand paving and adaptable to other paving units.

If using brick-in-sand paving, build temporary forms as if for a concrete footing (see pages 50–51). The forms should be set one brick length apart, around the base's perimeter, in a trench dug deep enough to allow for a 4-inch depth of concrete. Pour concrete (see page 37) and level it one brick thickness below the top of the forms with a bladed screed (see the drawing below; sand and paving bricks are not yet in place). Place edging bricks in the wet concrete and butt the joints; set bricks with a few taps of a rubber mallet.

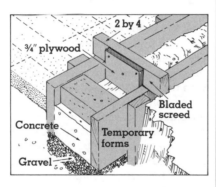

2 by 4

¾" plywood

Concrete

Bladed screed

Temporary forms

Gravel

Remove the forms the next day and allow the concrete to cure (see page 39) before packing soil around it. Use the bricks as a guide to screed the base's sand bed.

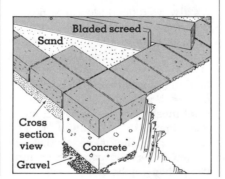

Bladed screed

Sand

Cross section view

Concrete

Gravel

Concrete edgings

Concrete edgings are built similarly to invisible edgings. In the drawing below, the poured concrete has been screeded flush with the top of the form boards (see page 37) so that it will be flush with the paving surface. Finish the concrete surface to suit your tastes, and allow it to cure (see page 39).

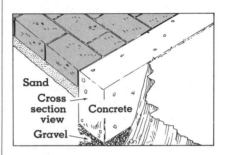

Sand

Cross section view

Concrete

Gravel

Wood edgings

To build wood edgings (also called headers or headerboards), choose a wood that is highly resistant to rot and termites—for example, pressure-treated lumber or the heartwood of cedar, redwood, and certain types of cypress. These woods will give you many years of service. The most common lumber size for straight edgings is the 2 by 4, though you can use thicker or thinner lumber if you prefer.

After grading, stretch a mason's line around the perimeter of the area to be paved. Drive 12-inch-long stakes (1 by 2s, 2 by 2s, or 1 by 3s) into the soil no more than 4 feet apart with their inner faces aligned with the mason's line. (To achieve the correct finished edging height, it may be necessary, before placing the stakes, to dig a narrow trench under the mason's line, so that the edging boards will be flush with the eventual paving surface.) Position the edging boards against the stakes and nail the stakes to the boards, holding the boards against the stakes with a heavy hammer or crowbar. Cut off the stake tops so that they angle up to the top of the edging boards. Pack soil tightly around the edging's outer perimeter.

Paving: Concrete

Concrete is the most versatile of all garden paving materials. Finished with plain or patterned textures, concrete complements all home styles.

Buying materials

You can purchase concrete for paving and other garden building projects in any one of four ways: bulk dry materials (sand, gravel, and cement); dry ready-mix, 1-cubic-yard trailerloads of ready-to-pour concrete; or truckloads of ready-to-pour transit-mix concrete.

Consider the size of your project, calculate the cubic feet or yards of concrete you'll need, and decide how much work you want to do before deciding which form of concrete is best suited to your project. Bulk dry materials are cheapest and most labor-intensive; dry ready-mix is expensive but more convenient.

Transit-mix is the best choice for a large-scale project because it enables you to finish it in a single pour, but you'll probably pay extra for delivery of quantities under 2 or 3 cubic yards, and cost may then outweigh convenience. There will be an additional charge if the truck has to wait after it arrives at the site. Special pumping equipment can be rented to reach awkward spots.

Building the forms

Forms are built the same way as wood edgings (see page 35). Wet concrete exerts a lot of pressure, so your forms should be strong and securely anchored to the ground. To prevent concrete from oozing under the form boards, pack soil against the forms around the outer perimeter of the area to be paved.

For 4-inch-thick paving (a common thickness for most garden paving), use 2 by 4s on edge for forms and 12-inch 1 by 2s or 2 by 2s for stakes. Nail the forms so that the paving surface will have at least ¼-inch slope per foot for runoff (see illustration below). Use double-headed nails so that you can remove them easily when you strip the forms. To make sure the forms are square, check with a steel square or with the method described on page 56.

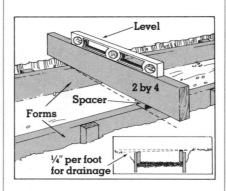

Level

2 by 4

Spacer

Forms

¼" per foot for drainage

If the area to be paved is large, you may find the concrete easier to pour and the finished appearance of the paving more attractive if you divide the area into modules. Build a permanent grid of 2 by 4 dividers as the outer perimeter's forms are built, and leave them in place after the concrete is poured. If you don't use permanent dividers in paving a large area, align temporary forms inside the outer forms so you can screed each batch of concrete as you pour it.

Reinforcing concrete paving

It's always a good idea to reinforce concrete paving with steel, particularly for large areas such as patios or when paving on unstable soil. The steel helps prevent cracking and will hold the pieces together if cracking occurs. Use 6-inch-square welded steel mesh sold by the roll at most home improvement centers; cut it to fit. Support the mesh on bricks, stones, or broken concrete so that it's midway in the slab.

Instead of steel reinforcement, you might include an expansion joint of compressible material every 10 feet or less, or you can pour the paving in modules formed by a grid of 2 by 4 dividers to allow for settling and for expansion and contraction of the paving.

If you're unsure about your need for reinforcement, consult your building department or a landscape architect.

Choosing a concrete formula

On page 52 you'll find a formula for making concrete from bulk dry materials. For most concrete building projects in the garden, this formula gives good results. Choose between the basic mix and one containing an air-entraining agent (which is usually needed in cold climates).

Refer to the table on page 52 to estimate how much of each bulk dry material you'll need.

Mixing concrete

An important factor in mixing concrete from bulk dry materials is the ratio of water to cement. Concrete hardens because the powderlike cement and water form an adhesive that binds the sand and gravel. Too much water thins or dilutes this adhesive paste and weakens its cementing qualities; too little makes it stiff and unworkable.

When working with small quantities, you can use a shovel or hoe to mix concrete on a wood platform or in a wheelbarrow. Measure the sand, spread it evenly, and spread the cement on top. Turn and mix the dry ingredients until no streaks of color appear. Add gravel or rock and mix until evenly distributed. Then make a depression in the mixture and slowly add water, turning the ingredients until they're thoroughly combined. Use a rolling motion with a shovel to speed the job.

For large quantities or if you're using an air-entraining agent, it's best to rent a power mixer.

How to pour concrete paving

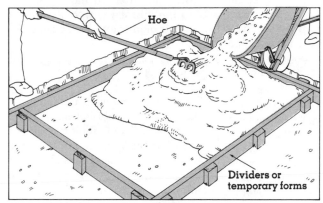

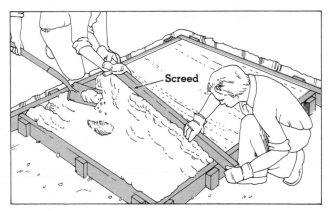

Pouring, spreading, and tamping. *Start pouring concrete at one end of form and spread with hoe or shovel. Tamp into all corners and against form. Do not overwork concrete.*

Screeding the concrete. *Move screed slowly along form, using rapid, zigzag sawing motion, to level concrete. Add or remove concrete as needed. On large pours, screed batch by batch.*

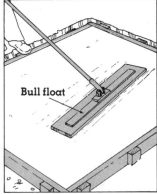

Edging and jointing. *Before and after floating (see right), run tip of trowel along edge and follow with edger. To reduce possibility of irregular cracks, use jointer to make control joint across concrete every 10 feet.*

Floating. *Use darby or bull float for large areas, wood float (see below left) for small areas, to smooth down high spots and fill hollows left after screeding. Add or remove concrete as needed. Overlap passes.*

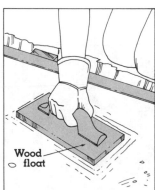

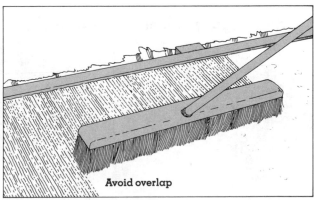

Final floating or troweling. *After water sheen disappears but before surface becomes really stiff, work with wood float for rough finish, with steel trowel for smooth. Cure as directed on page 39.*

Brooming. *For a nonskid surface, broom instead of floating or troweling. Texture depends on stiffness of bristles and whether they are wet or dry. Make a straight or wavy pattern. Cure as directed on page 39.*

Paving:
Exposed aggregate concrete

The pebbled texture of "seeded" or traditional exposed aggregate concrete makes it one of the most popular finishes for residential concrete paving. Sprinkled onto the poured concrete or exposed from within, the aggregate—your choice from many varieties of stone or pebbles—tempers concrete's mundane appearance. Its uniformly textured surface is a good choice for newly poured paths, patios, play areas, and poolside paving.

Another kind of exposed aggregate finish—but not suitable for heavily used areas—is cobbled concrete, made with cobblestones, river stones, or large pebbles set in freshly poured concrete.

Exposed aggregate finishes

Varicolored smooth gravel up to ¾ inch in diameter is the optimal choice for "seeded" and traditional exposed aggregate finishes. Prepare the base, build the forms, and pour the concrete (see pages 34 and 36–37).

For a *traditional exposed aggregate surface,* you expose aggregate that was mixed into the regular concrete formula before pouring. Simply finish the concrete through the wood float stage (see page 37). Don't overfloat, or you may force the aggregate too deep.

For a *seeded surface,* you purchase enough aggregate to cover the surface of your paving area with a single layer. After pouring, level the concrete ½ inch below the tops of the form boards with a bladed screed. With a shovel or by hand, evenly distribute the aggregate in a single layer over the concrete. Use a wood float or darby to press the aggregate down until it lies just below the surface of the concrete. Then refloat the concrete.

Periodically check the concrete—whether seeded or traditional—as it hardens; you won't be able to expose the aggregate if it becomes too hard. When the concrete can support your weight on knee boards without denting, you can expose the aggregate.

Starting at a corner on the high side of the concrete, wet the surface with a fine spray. Gently brush the surface with a broom (nylon bristles are best). Alternately spray and brush the surface, taking care not to dislodge the aggregate. Stop when the tops of the stones show.

Take extra care in curing exposed aggregate (see page 39)—the bond must be strong. You can re-move any cement haze left on the aggregate after curing by scrubbing the surface with a 10 percent muriatic acid solution (see page 41).

Cobbled concrete

Though time-consuming, finishing concrete with cobbled surfaces is relatively easy. Cobbled surfaces also offer the most flexible design potential because stones are hand set one by one. You can arrange stones in whimsical swirls, geometric patterns, or even mosaics.

Cobbled surfaces do have drawbacks, though. Stones are often slippery when wet, and their irregular shapes sometimes make walking uncomfortable.

The steps for finishing concrete with cobbled surfaces are very similar to those for exposed aggregate.

Prepare the base, build the forms, and pour the concrete (see pages 34 and 36–37). Don't fill the forms completely or the concrete will overflow when you press the stones into the mix (the larger the stones, the deeper the concrete must be and the more space for displacement you must allow). Finish the concrete through the floating stage. Place and firmly press each stone far enough into the concrete so the concrete grips the stone.

Cure the concrete (see page 39). To remove cement haze from the stones, scrub with a 10 percent muriatic acid solution (see page 41).

Seeded aggregate concrete paving

After screeding, *sprinkle the wet concrete with aggregate in a single layer, by hand or with a shovel, to ensure a smooth walking surface.*

Embed aggregate *by pressing with a wood float or darby until the concrete comes to the surface. Then, refloat the concrete.*

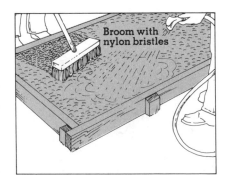

Wet the hardening concrete *with a fine spray; gently brush until the aggregate appears. Be careful not to dislodge the aggregate.*

Paving: Concrete blocks

Cast-concrete paving blocks are suitable for everything from pathways to patios. Laid on sand (according to the technique for bricks in sand—see page 40) or with large open joints (which are filled with soil in which turf or ground cover can be planted), they quickly become a part of your garden. You can buy the units precast, but why not consider making your own? The methods are simple, and the results distinctive.

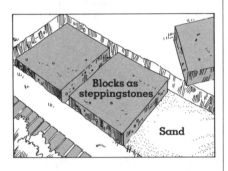

Two techniques for making your own blocks are discussed here. All the concrete finishing techniques presented on pages 37 and 38 apply to paving blocks as well—review them before you begin.

Another kind of purchased paving block (also illustrated) is the interlocking concrete paver, a European innovation available in a number of shapes and colors in many parts of the country. Laid in sand, the pavers fit together with closed joints. Some pavers have holes for planting—ideal if you want a grassy parking area.

Casting in-ground molds

The easiest way to make steppingstones is to pour them in place in the ground.

For each "stone," dig a hole 4

inches deep, shaping the hole as you choose. For easy walking, space the holes no more than 12 inches apart.

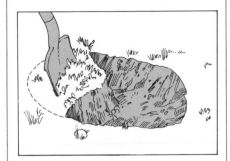

Use a mixture of 1 part cement, 2 parts sand, and 3 parts gravel (directions for mixing are on page 36) to fill the holes. Finish the tops with a trowel or wood float as described on page 37; then cover and cure the steppingstones as described at right.

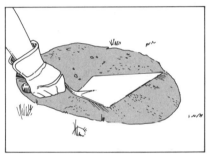

Casting in single molds

For a more uniform edge, you can cast your paving blocks in a mold.

A simple closed frame is the easiest to construct. You can make it from 2 by 4 lumber or any sturdy wood. For ease in unmolding, hinge two corners as shown in the illustration below, and add a hook and eye to close the open corner.

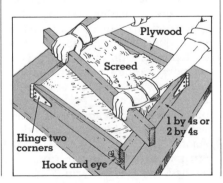

Here's how to cast your blocks: Oil the form with motor oil or a commercial releasing agent and place it on a smooth surface. Fill the box with a stiff concrete mix, packing it in. Then screed the top with a straight board. You can float the surface, if you like (see page 37), or use the block's smooth underside for the stepping surface.

Wait until the concrete has set for a few hours before unmolding, then cure the blocks (see below).

Curing concrete

To cure concrete, cover the exposed surfaces with plastic sheet or a commercial curing agent. You can use burlap or straw, but you'll have to keep it damp throughout the curing process. Then allow the concrete to dry for at least 4 days before uncovering (the cover is especially important in hot, dry climates, to slow the drying process).

Setting interlocking pavers

You'll have less work to do if you set these pavers in areas shaped in squares or rectangles. To minimize cutting, design the area to use full pavers. Use one of the edgings described on page 35 to hold the outer pavers in place, or plant turf to the same height as the finished surface.

Set the pavers on a 2-inch sand bed as for brick (see page 40). The distance between the top of the bed and the finished grade should equal the thickness of the pavers. Settle the pavers into the sand bed with several passes of a roller or small-plate vibrator, which you can rent.

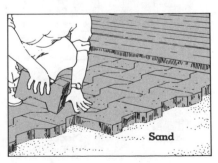

Paving: Brick

For a path or patio with a gracious look, brick is a good choice. Its simple form lends itself to a variety of paving patterns, and you'll find bricks in colors and textures compatible with almost any style of home.

Though many types of brick can be used, common brick is the favorite for most paving, both for its slip-resistant texture and nonglare surface—and it's usually the least expensive. The color of common brick ranges from warm brown to soft red, depending on the area of the country in which it is made.

Illustrated below are a few examples of the dozens of bonds or patterns you can choose from. Some patterns are complex, requiring a good deal of accuracy; with others, a large number of bricks must be cut. Because brick sizes vary, the suggested ½-inch joint for grout may not suit your choice of bond. Check with your building supplier for correct joint width.

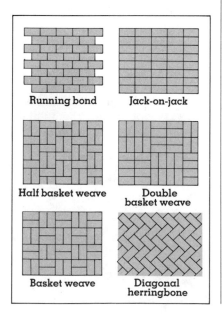

Running bond | Jack-on-jack

Half basket weave | Double basket weave

Basket weave | Diagonal herringbone

Cutting brick

Soft common bricks are fairly easy to cut with a brickset. Place the brick on sand or soil. Using the brickset and a club or other soft-headed hammer, score the brick on all sides where you want to make the cut. Then cut the brick with a sharp blow to the brickset. Always wear safety glasses.

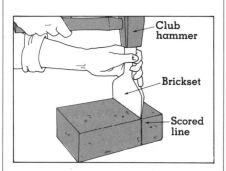

Club hammer

Brickset

Scored line

If you're using especially hard bricks, or want very precise brick-work, you'll be wise to rent a masonry saw. You can buy a masonry-cutting blade to use with your own saw, but brick dust can damage the bearings.

Setting brick in sand

Setting brick in sand is the easiest method for the beginner. Before you begin, you'll need to grade the paving site, construct edgings (see page 35), and lay a 2-inch-deep sand bed. Don't skimp when ordering the sand; and do have it dumped next to your work area.

Screed the base. If your project is a brick walk or something similarly narrow, use the edgings as guides for the bladed screed. When paving large areas, use one edging as a guide and set a 2 by 6 on edge on the soil about 3 feet from the edging, with its top level with the edging, as a temporary guide (see illustration following). The depth of the screed's blade should equal the thickness of the brick. Place dampened sand between the guides and screed it smooth. Tamp or roll the sand, spray with water, add more sand as needed, and screed again.

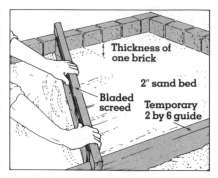

Thickness of one brick

2″ sand bed

Bladed screed

Temporary 2 by 6 guide

Set the bricks. Working outward from one corner, lower the bricks into position butted up against each other (don't slide them—you'll trap sand between the bricks) and bed them by tapping them with a rubber mallet, the handle of a regular mallet, or a block of wood. Stretch a mason's line, as shown, to aid alignment. If you're laying brick in a large area, move the temporary 2 by 6 guide as you work. Add dampened sand; then screed, using the newly laid brick as the other guide.

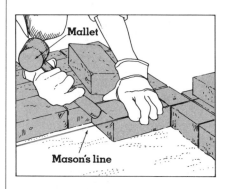

Mallet

Mason's line

Sand the joints. Spread fine sand over the finished paving surface. Let it dry thoroughly; then sweep it into the joints, adding sand as necessary to fill them. Finally, wet the finished paving with a fine spray.

Fine sand

Setting brick with dry mortar

If you want to set your bricks in sand but prefer a wide joint, plan to grout the joints to keep the bricks from shifting. (Should the sand bed ever settle, you'll have to regrout some joints.) Before you set the bricks, prepare the edgings and the sand bed as described on the facing page; then follow the steps below.

Place bricks and mortar. Set all the bricks on a 2-inch sand bed, leaving ½-inch-wide joints (use a piece of ½-inch plywood for a spacer and a mason's line for alignment). Check frequently with a level. Prepare a dry mortar mix of 1 part cement and 4 parts sand; spread the mix over the surface, brushing it into the open joints. To avoid disturbing the bricks, kneel on a piece of plywood.

Mortar mix

Tamp the mortar. Use a piece of ½-inch plywood to tamp the dry mix firmly into the joints; this improves the bond. Add more mix if needed. Carefully sweep and dust the tops of the bricks before going on to the next step, because any mix that remains may leave stains. (Some staining, though, is unavoidable.)

Wood tamper

Wet the surface. Using an extremely fine spray, wet the paving. Don't allow pools to form and don't splash the mortar out of the joints. During the next 2 to 3 hours, periodically wet the paving to keep it damp. Tool (finish) the joints (see "Setting brick in wet mortar," below) when the mortar is firm enough. For information on removing excess mortar from the bricks, see "Grout the joints," following.

Setting brick in wet mortar

Bricks wet-mortared over a concrete slab or other solid base make a firm, durable paving (see pages 36 and 37 for information on poured concrete). The bricks can be laid on any properly prepared base, including existing concrete in good condition. Wet your bricks several hours before you start work, to prevent them from drawing water out of the mortar.

Prepare the bed. Lay and screed a ½-inch-thick wet-mortar bed between edgings placed against the slab (see page 35 for edgings). Use a mixture of 1 part cement and 4 parts plastering sand, gradually adding water so the mortar spreads easily but doesn't run.

Use a bladed screed that rides on the edgings, extending one brick thickness below them; set the edgings for this depth plus ½ inch for the mortar. (If your area is large, you may need to use temporary guides as described in "Setting brick in sand" on the facing page.) Mix only as much mortar as you can use in an hour or so, and screed only about 10 square feet at a time.

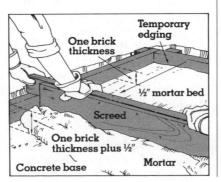

Temporary edging
One brick thickness
½" mortar bed
Screed
One brick thickness plus ½"
Concrete base
Mortar

Set the bricks. Place the bricks in the pattern of your choice, leaving ½-inch-wide joints between them (use a piece of ½-inch plywood for a spacer and a mason's line for alignment). Gently tap each brick with a rubber mallet to set and level it.

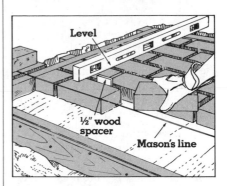

Level
½" wood spacer
Mason's line

Grout the joints. Use a small trowel to pack mortar (1 part cement to 4 parts sand, plus ½ part fire clay to improve workability) into the joints, working carefully to keep mortar off brick faces. Tool (finish) the joints with a jointer (illustrated below). When grout is finger hard, scrub paving with damp burlap to remove stains. Keep grout damp for about 24 hours by covering the paving with plastic sheet, and stay off the paved area for 3 days.

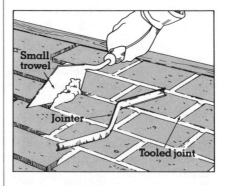

Small trowel
Jointer
Tooled joint

Cleaning brick

To remove dried mortar, use a solution of 1 part muriatic acid added to 9 parts water. Soak the area to be cleaned with water. After the water settles in, brush 10-square-foot areas with the acid solution, using a stiff brush. (Wear gloves and safety glasses, and use a plastic bucket.) Rinse thoroughly and immediately with a hose to prevent acid stains.

Ideas for garden steps

On the facing page are ideas for steps featuring six different paving materials you can adapt to meet your particular needs.

Consider railroad ties or flat rocks for their rustic appeal, or brick for its range of effects, from rustic to formal. Poured concrete is excellent for utility and can be given an attractive exposed-aggregate finish. Concrete blocks make simple, inexpensive steps, or they can serve as a foundation for other materials. (For information on wood steps, see page 72.)

Designing the steps

To design steps, you must determine the proportions of the steps and the degree of slope.

Proportions. Architects and builders long ago worked out a set of ideal proportions for steps: twice the riser height (each step's vertical dimension) plus the tread depth (each step's front-to-back horizontal dimension) should equal 25 to 27 inches. A comfortable average is a 6-inch riser with a 15-inch tread. Risers should be between 4 and 7 inches high; treads should be at least 11 inches deep (see illustration below).

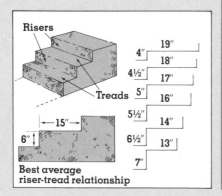

4"	19"	
4½"	18"	
5"	17"	
5½"	16"	
6½"	14"	
	13"	
	7"	

Risers

Treads

←15"→

6"

Best average riser-tread relationship

Dimensions. Calculate the change in level of your slope, using the simple method shown in the illustration below. The distance from A to B is the change in level (also known as the "rise"); the distance from A to C is the "run"—the minimum distance your steps will run.

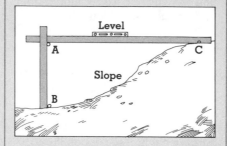

Level

A   C

Slope

B

Determine the number of steps you'll need by dividing the desired riser height into the total rise of the slope. Check the illustration under "Proportions" to see if the corresponding minimum tread will fit into the slope's total run. You will probably have to excavate—cut into the slope—or fill under each step with well-compacted soil to make the slope fit the stairway.

Plan on a minimum width of 2 feet for simple utility steps. Four feet is a good minimum width for more gracious steps—and make that 5 feet if you want two people to be able to walk abreast.

You don't have to build steps in a single flight straight up the slope. You can break them with a landing, let them ascend in a gentle S-curve, or use a combination of the two.

Planning. Take time to make a scale drawing before you begin—you don't have to live with a bad drawing, but a project gone awry is a different matter. Plans for steps that abut a sidewalk or other public access should be cleared with your building department.

Preparing the base

First, shape your slope to form rough steps in the earth, keeping the treads as level and the risers as perpendicular as possible, and allowing space for the thickness of the setting bed (if one is used) and the paving units.

The next step—laying the setting bed—depends on the material you use for the steps. Flagstones, large stones, railroad ties, concrete blocks, and wood rounds can be set directly on well-tamped soil. Brick, concrete blocks, tile, and concrete paving blocks can be laid on sand that is held in place by wood risers (see illustration for "Concrete pavers" on facing page).

Poured concrete steps or a concrete base for steps built of brick or concrete paving units require a gravel base and wood forms (see the instructions below). Be sure to allow for the 6-inch minimum thickness of the gravel setting bed and the 4-inch minimum thickness of the concrete, on both treads and risers, when you form the earthen steps in your slope.

Building forms for concrete steps

Use 2-inch-thick lumber to build forms for concrete steps or for a concrete base for brick or concrete paving units. The forms are very similar to the wood edgings described on page 35; position them so that the concrete will be 4 inches thick on both treads and risers (see illustration below). Level and nail the forms to stakes driven into the ground on the outside of the forms so that you can remove both forms and stakes after the concrete has cured (see page 39).

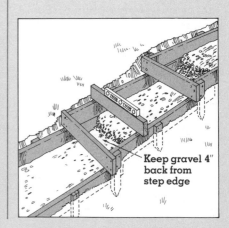

Keep gravel 4" back from step edge

Six types of step construction

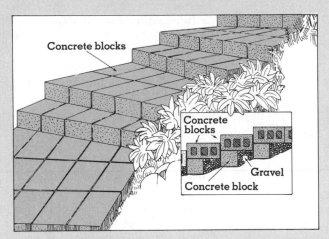

Concrete corner blocks set on their sides are supported by concrete blocks and gravel.

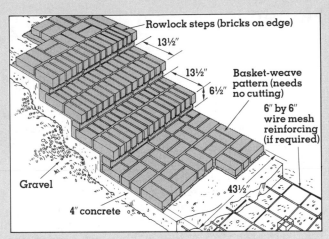

Brick treads and risers are set in mortar, supported on concrete base.

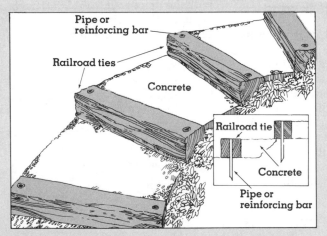

Railroad tie risers, held in place with pipe or bars, retain concrete treads.

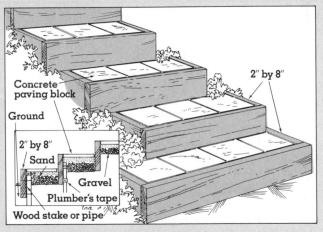

Concrete pavers are set on sand held in place with wood risers. Same method can be used for brick or tile.

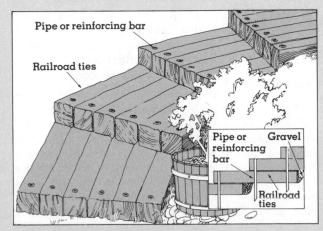

Railroad ties, set directly on excavated ground, are held in place by their own weight and by pipes or bars.

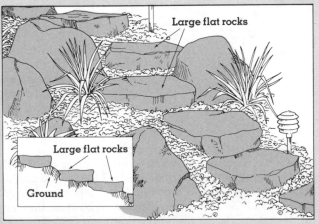

Large flat rocks, set directly on excavated ground, require lots of helpers with strong backs to place.

Paving: Flagstone

Flagstone makes an attractive, durable surface for patios and walks; properly constructed, such paving can last almost forever. For many people this lasting quality outweighs the fact that flagstone is a relatively expensive paving material. Choosing a stone quarried close to your home will save shipping charges.

Fitting and cutting flagstone

Most flagstones are irregularly shaped; you'll need to fit and cut each piece before setting it. After you've prepared the area that's to be paved (see "Preparing a base," page 34), lay out all the flagstones on the surface and then shift the stones around until you achieve the most pleasing design.

When a stone requires cutting or trimming, let the adjoining stone overlap it. Mark the cutting line with a pencil, using the edge of the top stone as a guide.

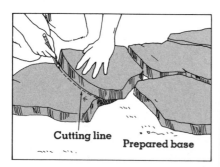

Cutting line   Prepared base

Then cut the stone by scoring a ⅛-inch-deep groove along the line with a brickset or stonemason's chisel. Place a length of wood under the stone so that the waste portion and a slight portion of the scored line overhang it. Strike sharply along the line with a brickset and club or other soft-headed hammer. Be sure to wear safety glasses.

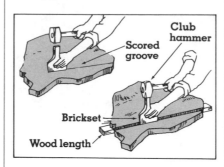

Club hammer
Scored groove
Brickset
Wood length

You can trim the stone and cut curves by chipping off pieces of the waste portion with a club or other soft-headed steel hammer until you achieve the desired shape.

Setting flagstone in sand

This is the easiest method for setting flagstone. Begin by preparing a 2-inch-deep sand bed, following the directions for brick (see page 40). Set the stones and level them, scooping out or filling in the sand bed under each flagstone as needed. Plant the open joints with grass or a ground cover that will fill in the gaps, yet survive being walked on; or fill the joints with sand or soil.

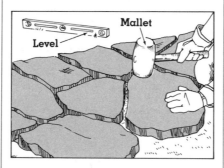

Mallet
Level

Setting flagstone in wet mortar

For the most permanent flagstone surface, set stones in a mortar bed on top of a 3-inch or thicker concrete slab. You can use an existing slab (clean and in good condition) or one built for this purpose (see pages 36 and 37 for information; a new slab should cure for at least 24 hours before you lay the stones).

Arrange and fit the stones in the desired pattern before laying the mortar setting bed. If they're porous, you'll need to wet them a few hours before you set them in mortar.

Ask your stone dealer if you should apply a bonding agent to the slab or mix it into the mortar.

Lay the stones. Mix a batch of mortar (3 parts sand to 1 part cement) to cover 10 to 12 square feet. Add water slowly—to support the weight of the stones, the mortar should be stiff.

Starting at one corner, remove a section of stones and set them to one side in the same relative positions. With a trowel, spread enough mortar (at least 1 inch deep for the thickest stones) to make a full bed for one or two stones. Furrow the mortar bed with the tip of your trowel. Set each stone firmly in place and bed it by tapping with a rubber mallet.

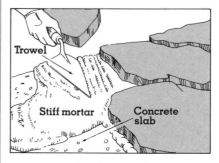

Trowel
Stiff mortar
Concrete slab

To maintain an even surface, use a level. If a stone is not level, lift it and scoop out or add mortar as needed. Again, furrow the mortar and bed and level the stones.

Align the edges of the outer stones with the perimeter of the slab, or let them overhang it slightly. Remove any excess mortar from the perimeter after bedding the stones.

Grout the joints. After the mortar has set for 24 hours, grout the joints with the same mix of mortar used for the bed, plus ½ part fire clay to improve workability. Using a trowel, pack the joints with mortar, then smooth the joints, also with a trowel. Clean the face of the flagstones with a sponge and water as you work. Keep the grout damp for the first day by sprinkling or covering with plastic sheet, and keep off the area for 3 days.

Paving: Tile

Though tile is expensive, it ranks high in durability and stain-resistance. Outdoor tile is usually unglazed, taking its color from the clay—gray to brick red; thickness is usually ⅜ to 1 inch. Unless a special grit has been added to glazed tile, it may be slippery when wet.

To minimize tile cutting, try to plan your area to accommodate full tiles. If you cut tiles, use a wet-saw.

Setting tile in sand

You can lay ¾-inch-thick tile in sand. Follow the directions for "Setting brick in sand" (page 40), but use only a ½-inch sand base; a thicker base may cause the tiles to tilt out of position. Use butted joints for added stability.

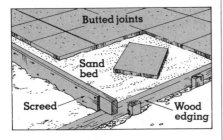

Setting tile in dry mortar

To set tile in dry mortar, you need a specially prepared base. Level the soil 3 inches below the desired grade of the paved surface and build temporary wood edgings as guides around the perimeter of the area to be paved (see the information on grading and edgings on pages 34 and 35). Set the top surfaces of the guides so they will be flush with the bottom of the tile. Pack soil around the outer perimeter of the guides.

Fill the area with sand. Using a straight board, screed it level with the tops of the temporary guides. Over every 100 square feet of surface, evenly distribute 2 bags of dry cement. Mix sand and cement together with a rake, being careful not to mix in the earth below; then tamp and rescreed the mix to a smooth surface flush with the guide tops.

Sprinkle the screeded sand-cement mixture evenly with ½ bag of dry cement for every 100 square feet. Place the tiles on the surface, aligning them with ⅜-inch open joints as shown (use a length of ⅜-inch plywood as a spacer). Make sure tiles are level, and bed them by tapping the centers with a rubber mallet.

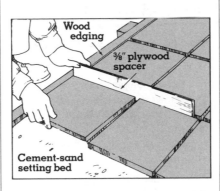

Wet the entire area with a fine spray until sand, cement, and tile are moistened, but don't let the water puddle or splash mortar out of the joints. Allow the tiles to set for 24 hours; then grout and tool the joints with a mixture of 1 part cement and 3 parts sand as described in the following section. After the grout has cured, remove the temporary edgings and fill in the recesses with soil.

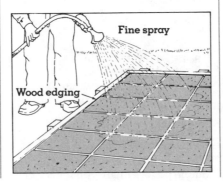

Setting tile in wet mortar

The illustration below shows a particularly stable method of laying tile: in a 1-inch mortar bed over a newly poured or existing concrete slab (see pages 36 and 37 for information on poured concrete), using the same basic method as for brick (see page 41). The edgings shown can be left in place or removed after the mortar has set. The illustration shows a bladed screed (made by nailing a piece of ¾-inch plywood to a 2 by 4) that rides on the edgings and levels a 1-inch mortar bed; when tiles are laid on top, they will be flush with the tops of the edgings.

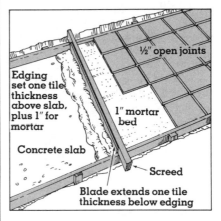

Allow the tiles to set for 24 hours; then grout them with mortar (1 part cement and 3 parts sand). The mortar should be just thin enough to pour. Use a bent coffee can to fill the joints, cleaning away spills immediately with a damp sponge. Tool the joints as you would for brick (see page 41), using pipe or a jointer. Keep the grout damp for the first day by sprinkling or covering with plastic sheet, and stay off the paved area for 3 days.

Paving: Adobe

It's difficult to equal adobe paving—common in the American Southwest—for friendly, rustic charm and historic elegance. Spaced with wide joints, adobe pavers make an excellent base for a living floor of moss or creeping plants that grow to fill the joints and soften the look of the surface.

Today's adobe is made with an asphalt emulsion that coats each particle of soil and makes the adobe virtually waterproof. Adobe pavers, 2½ inches thick and measuring 6 by 12 or 12 by 12 inches, are made with an extra dose of emulsion for additional durability.

Adobe can be less expensive per square foot than brick. Asphalt-stabilized adobe is suitable for use in any part of the country, but most adobe manufacturing facilities are in the West and shipping charges may make it an uneconomical choice in other regions.

Making your own adobe

Areas of the country other than the West have soils suitable for adobe pavers, so you might consider making your own if you find it too expensive to buy them.

Though the method for making adobe pavers is simple, it requires a lot of labor. You begin by combining soil and water with an asphalt emulsion mixture of road oil and soap until the mixture is stiff yet creamy, loading it into a simple wood mold, and screeding off the surface with a board (see illustration following). Then you remove the mold and allow the adobe pavers to cure in the sun. Finally, after a day, you stand the pavers on end and leave them there to cure for 3 weeks.

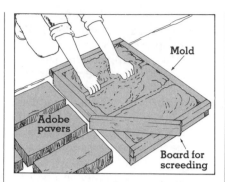

Mold
Adobe pavers
Board for screeding

You'll find more information on making your own adobe blocks in the *Sunset* book *Basic Masonry Illustrated.* You can obtain detailed instructions on how to make adobe by writing to International Institute of Housing Technology, % Executive Vice President, California State University, Fresno, CA 93726.

Setting adobe pavers

You set adobe pavers on a sand bed as you would brick (see page 40). Though you can set them with butted joints, you'll find open joints easier because they compensate for size irregularities in the pavers.

Edgings for adobe paving. If your adobe is flush with the ground, the soil can be packed against the edges of the paved area to retain the pavers. If the top of the finished paving will be above the ground, build one of the edgings shown below.

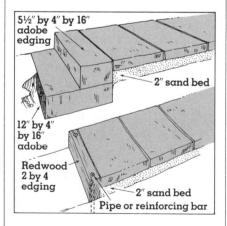

5½" by 4" by 16" adobe edging

2" sand bed

12" by 4" by 16" adobe

Redwood 2 by 4 edging

2" sand bed
Pipe or reinforcing bar

Making the sand bed. Lay a 2-inch-deep sand bed as you would for brick (see page 40). If you're setting the adobe pavers flush with the ground,

use temporary wood edgings as screeding guides (see page 35).

Setting the adobe. Set rectangular or square adobe pavers in running bond or jack-on-jack, and rectangular pavers in basket-weave pattern (see page 40), using open joints.

Starting at one corner, lay a paver in place; make sure it doesn't sit on a hump or straddle a hollow—either one can lead to cracking. Scoop out or add sand as needed to set the paver. Continue setting more pavers, using a wood block to space the joints evenly and a mason's line to keep them in alignment. Check each paver with a level.

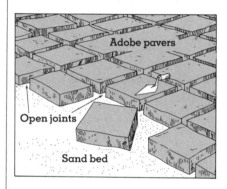

Adobe pavers
Open joints
Sand bed

You won't have to cut any adobe pavers when you use the jack-on-jack or basket-weave pattern, if you accurately plan the area to use whole pavers. If you use a running bond, cut the pavers as you would brick (see page 40) or cut them with an old carpenter's saw.

Finishing the joints. The best way is to fill the joints with sand or soil, or plant the crevices with moss or creeping plants.

If you like the clean look of mortared joints, you can grout the joints with a mixture of 1 part cement and 3 parts sand, adding 1½ gallons of asphalt emulsion (available from your adobe paver supplier) per bag of cement as a water seal. Follow the directions for grouting brick under "Setting brick in wet mortar," page 41. But beware: mortared joints can be a problem. Manufacturers have found that many adobe failures, aside from wear and tear, start at such joints and are caused by wetting and drying.

Paving:
Quick & easy

If you don't want to invest time, money, and effort in a permanent paving, consider these quick, inexpensive, and easy alternatives. You can create pavings with loose wood or rock products, or pave an area with wood rounds, blocks, or lumber.

Paving with loose materials

The ideal "soft" or "loose" paving material—bark, gravel, decomposed granite, crushed rock, and redrock—stays where it's laid, drains well so it can be walked on when wet, and keeps shoes free from dust and mud. It is sold by the cubic yard or by the ton.

Grade the area to be paved and build wood or masonry edgings (see pages 34 and 35) to contain loose pavings, preventing them from spreading. For large areas, a grid of 2 by 4 dividers, built as wood edgings are built, keeps the paving material more uniformly distributed.

Bark. By-products of the lumber industry, barks vary in size, so choose the one that gives the effect you want. After grading and edging, and before laying bark, you may want to sterilize the soil to minimize weed growth; if so, consult a nursery so as not to jeopardize nearby plantings. Lay bark 2 to 4 inches thick in well-drained areas. Puddled water encourages rot and insect infestation.

Gravel. Though gravel drains well, it needs frequent raking to maintain a neat appearance, and it can be difficult to walk on. Gravels range from ⅜ inch (pea gravel) to ¾ inch in diameter or larger.

One method of building a gravel paving resists the natural tendency of gravel to move underfoot and has the added advantage of using less expensive gravels as a base.

After grading and edging, lay untreated road base (sharp-edged crushed rock ranging in size from sand to ¾ inch or larger in diameter) 2½ to 4 inches thick. Wet it down and roll it. If you're using a hand roller, you'll get better compaction by laying the base in 1-inch layers, wetting and rolling each layer. While the base is still wet, add a ¾-inch layer of pea gravel, colored rock, or other gravel and roll it into the base.

Decomposed granite and crushed rock. Both materials make good pavings because the particle sizes range from sand to quite large pieces of rock so they pack tightly together and don't move underfoot. After grading and edging, follow the procedure for untreated road base as described under "Gravel," above. In this case, however, the ¾-inch topping layer is optional.

Redrock. The actual color of this material varies from yellow to brown to red, depending on the region. It is also sold under several names. A mixture of clay, sand, and soil, redrock compacts well when damp. It does break down, becoming dusty in time, so periodically, you'll have to add new material.

Because redrock is not graded by size, you'll need a ¼-inch mesh sieve to sift out enough fine material to set aside for a ¾-inch-thick topping. Then, after grading and edging, lay and level a 2½ to 4-inch-thick base of the larger material. Dampen and roll the material, as described for gravel. Finally, distribute the finer material evenly over the base, dampen, and roll it.

Paving with wood

Because wood decomposes eventually and can be attacked by insects, wood paving is less permanent than masonry. By using rot and insect-resistant woods (heart redwood, cedar, and cypress) or pressure-treated lumber, you can slow the decomposition.

Wood paving can be made from rounds (slabs cut crosswise from a log) set on a gravel bed; blocks cut from railroad ties or other large timbers and set in sand as for brick (see page 40); modular sections built with 2 by 4s or 2 by 6s; and long 2 by 4 planks nailed on 4 by 4 wooden sleepers. All are shown below.

Types of wood paving

Rounds

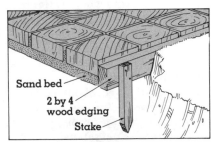

Sand bed
2 by 4 wood edging
Stake

Blocks

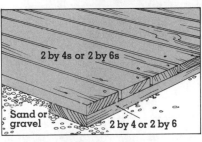

2 by 4s or 2 by 6s
Sand or gravel
2 by 4 or 2 by 6

Modular sections

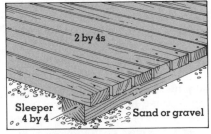

2 by 4s
Sleeper 4 by 4
Sand or gravel

Planks on sleepers

Avoiding drainage problems

Good drainage is the cornerstone of lasting gardens and patios. Without it, prized pavings and precious plants can drown in puddles or ponds during the wet season. Worse still are the scars that rain and melting snow leave on unprotected slopes.

Whenever a house or garden structure is built on a piece of ground, the ability of that piece of ground to absorb and gently drain away a downpour is greatly reduced. Roofs and patios do not retain water; they allow it to run off immediately. The excavating, filling, and leveling done during the construction process alter the ground's contours, with the result that puddles and washouts may occur even in gentle rains.

Most of these drainage problems are preventable by proper grading (see illustration below) during the original home construction. Even so, during the first year after a home's construction, you should keep a close watch on how and where water collects and drains away. If problems emerge, try solving them by correcting the grade, by using one or more of the methods illustrated on the facing page, or by consulting a landscape architect about major reshaping. Be careful that you don't solve your drainage problems at the expense of your neighbors.

If your house is several years old and your garden is fairly well established, drainage probably is under control. Established plants and lawns hold water and soil in place, and they transpire a considerable amount of excess rain and snow melt into the atmosphere. But even a fully landscaped home may have drainage problems, and you may create more when you build any new garden structures or relandscape.

Grading

A house can be swamped with water problems—from a leaky foundation to a flooded basement—if its site isn't graded for good drainage. Gardens and patios are equally susceptible to water's destructive force. When you're planning garden and patio projects, it's necessary to figure your grading needs.

The object of grading is to protect structures and landscaping from rain, melting snow, and even water used to irrigate planted areas. You accomplish this by sloping the land away from the area you want to protect. In fact, all horizontal surfaces in the garden, with the possible exception of decks, should be sloped (about ¼ inch per foot) to guide the water in the direction you want it to go and to keep puddles from forming.

For many homeowners, grading may involve nothing more than leveling humps, filling depressions, and smoothing out the ground to provide a gentle slope and a swale to carry excess water away from all structures and into the street, storm drain, or other area (see illustration below). If this is true in your situation, see page 34.

The grading of some properties may be more complicated, however. Low-lying areas, steep slopes and hillsides, and areas with unstable soil can pose serious grading problems. In these cases, a landscape architect can be a great help; often such a professional is essential.

If you're planning a major reshaping of your property, the help of a landscape contractor with earth-moving equipment will save you a lot of sore muscles and blisters, as well as time.

Other solutions

Grading alone may not be enough to rid your property of drainage problems. You may have to attack them by using one or more of the methods shown on the facing page—drainage ditches, baffles, dry wells, catch basins, terraces, or riprap (embedded stones).

These garden drainage projects are effective because they slow the speed of surface water runoff or provide a channel to direct water elsewhere.

If you anticipate complex drainage problems, or if you live in an area where the ground freezes, check with your building department or a landscape architect for the best way to assure a stable site. Either can advise you on which of the following ideas will best solve your problems.

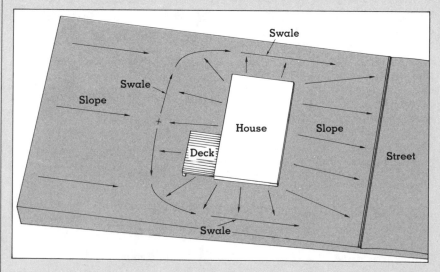

Grading. *Improve drainage by grading to lead water away from structures, into swale or wide ditch, and out to street.*

Six ways to improve drainage

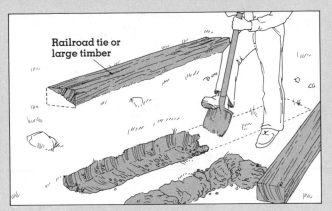

Baffle. Railroad tie or large timber set across slope and partly in ground will reduce runoff. Start with a few baffles and watch results. Add baffles until runoff is controlled.

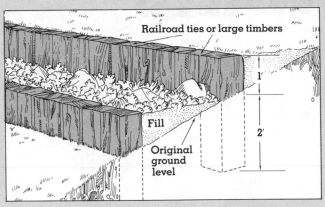

Terraces. To slow runoff on slope, build a series of low walls of large timbers (illustrated above), wood, masonry, or stone. Work from the bottom of the slope up, building each wall and filling in behind it.

Riprap. Cover a slope with stones at least 6 inches in diameter to slow runoff. Half the diameter of each stone should be embedded in ground. The steeper the slope, the closer together stones should be.

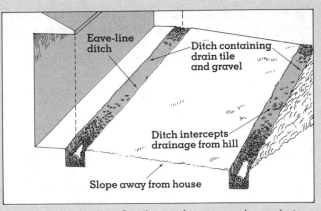

Intercepting ditches. On sloping lot, use perforated pipe or drain tile in gravel-filled trench, about 1-foot square, to collect water at bottom of slope or off roof and carry it away.

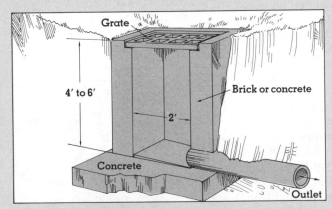

Catch basin. To drain water from low-lying area, build a catch basin at lowest point. Run pipe from sloped bottom of basin to disposal area such as street or storm drain. Cover top of basin with iron grill flush with ground.

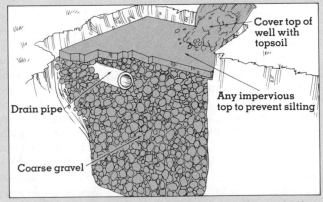

Dry wells. To dispose of water that can't be drained off, dig a dry well down to gravel or sandy soil. Bottom of hole must be above water table. Fill hole with coarse gravel, install drainpipe to lead water into it, and cap.

WALLS

Barriers of concrete, brick, stone & adobe

Masonry is your best choice when you want a wall that will last—one you can build, then admire for a lifetime.

In this chapter, you'll find easy-to-follow instructions on how to build walls of poured concrete, brick, concrete block, stone, and adobe block. The building techniques could apply equally to such projects as planters, borders, hearths, and barbecues—once you've mastered the skills needed to build walls with each material, you'll be able to strike off on your own.

Many municipalities impose restrictions on the building of freestanding masonry walls more than 3 feet high. You'll need a building permit or variance for such a wall, and you may need to have it designed by an engineer. For these reasons, the discussion that follows applies only to walls up to 3 feet high. Check with your local building department if you're planning a higher wall.

Walls:
Building a footing

Except for a dry stone wall, always use concrete for the footing of a masonry wall. Typical footings are twice the width of the wall and at least as deep as the wall is wide, but be sure to consult local building codes for exceptions to this rule of thumb. Your building department will also supply you with any information you need on the requirements for steel reinforcing.

Lay all footings on a gravel bed, usually about 6 inches thick, so that the bottom of the footing is below the frost line. The underlying earth should be firm and, ideally, of uniform consistency. If necessary, tamp the bottom of the trench. Though the trench bottom should be as level as possible, don't fill in any low spots; the gravel will take care of that.

To build a footing, simply dig a trench to the desired dimensions of the footing, add a gravel base, and pour concrete into the trench (see page 52 for information on concrete). In cases where the earth is too soft or too damp to hold a vertical edge, you can build a form to hold it, using the technique for wood edgings discussed on page 35. An alternative braced form is shown on the facing page ("Unkeyed form").

The footing should be flat on top if a brick, adobe block, concrete block, or stone wall is to be built on top of it. If you plan to pour a concrete wall in place on top, you have two choices: you can pour the footing and wall separately, or you can pour them at the same time.

Most walls will require a separate pour in a sturdy form and should be keyed into the cured footing as shown in the illustration following. (The form for a keyed footing is illustrated on the facing page.)

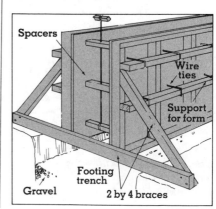

For some lower, lighter walls, your building department may allow you to pour the wall at the same time you pour the footing (see below).

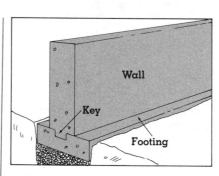

How to pour a footing

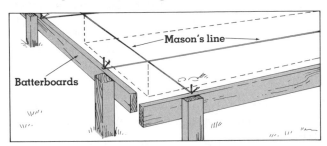

Lay out the footing. *Build batterboards of equal height at each corner. Stretch mason's line between opposite batterboards to mark outline of trench. After digging trench, replace line to use as guide for forms.*

Prepare the base. *Level bottom of trench and tamp until firm. Place 6-inch layer of gravel in trench and then level; top should be below frost line in your area. On sloping sites, use stepped footing shown below.*

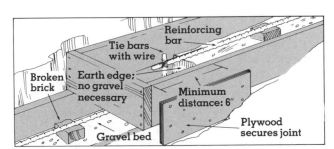

Unkeyed form for soft earth situations. *Build form along both sides of trench. Use stakes and braces to hold forms in place, making sure forms are level. Support any reinforcing bars on broken bricks set on top of gravel.*

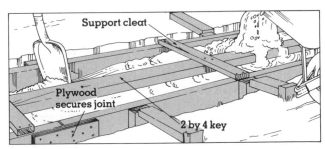

Keyed form for footing of separately poured wall. *Use 2-inch-thick lumber (held in place with stakes and braces) for forms, making sure tops are level. Hold 2 by 4 key in place by nailing to support cleats.*

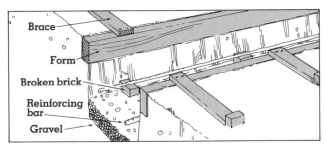

Form for stepped footing. *On slopes, build stepped forms and secure form sections to each other by nailing a piece of plywood across each joint. Tie any required reinforcing bars with wire as shown.*

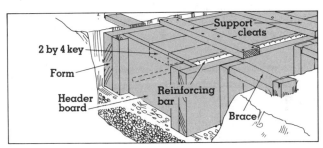

Pour and tamp concrete. *Oil forms with motor oil and pour concrete, tamping it firmly with shovel. Run shovel up and down along forms to eliminate voids. Work from one end; to allow for screeding, don't overfill form.*

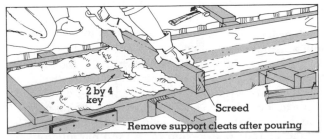

Screed concrete. *Use a piece of wood to screed concrete level with top of forms. Zigzag screed from one end to the other, striking off high spots and filling hollows. Insert any required vertical reinforcing bars into wet concrete.*

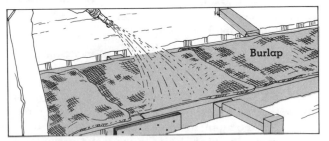

Cure concrete. *After screeding footing, cover with plastic sheet, burlap, or newspapers. Keep porous coverings wet by spraying them with water several times a day. Cure concrete for 4 days.*

Walls: Poured concrete

Poured concrete is today's answer to stone—with it, you can build a wall of almost any shape or size. Your building department may require you to reinforce the concrete with steel, and you should check for this requirement as well as for any others before starting your project.

Choosing a concrete formula

For most residential wall and paving projects, the concrete formula that follows will give good results. Choose between the basic mix and one containing an air-entraining agent to help prevent cracking in cold climates. (If your project is large, consider ready-to-pour transit-mix, described on page 36.)

Basic concrete. Use this formula for regular concrete, if you're mixing it from bulk dry materials. All proportions are by volume.

- 1   part cement
- 2½   parts sand
- 2¾   parts gravel
- ½   part water (approximate, depending on wetness of sand)

The sand should be clean river sand (never use beach sand); the gravel diameter should range from quite small to about ¾ inch. The water should be drinkable—neither excessively alkaline nor acidic, nor containing organic matter.

For information on mixing concrete by hand, see page 36.

Air-entrained concrete. Adding an air-entraining agent to the concrete mix creates tiny air bubbles in the finished concrete. These help it to expand and contract without cracking, a quality important in areas with severe freeze-thaw cycles. The agent makes concrete more workable and easier to place—the extra workability means you can add less water to a batch. Within limits, this makes the finished concrete stronger. An air-entraining agent is routinely added to transit-mix, whatever the local climate. You can add an air-entraining agent to your own concrete mix when you're mixing it by machine.

The amount of agent needed varies by brand. Your supplier can advise you on this and tell you about any other adjustments to the basic formula that may be necessary.

How much to buy. Refer to the table below to estimate what you'll need. It's a good idea to figure about 10 percent extra—you'll waste some, and you don't want to run short.

Here's what you'll need for each 10 cubic feet of finished concrete:

Separate ingredients:

Cement	2.4 sacks
Sand	5.2 cubic feet
Gravel	7.2 cubic feet
Ready mix:	.37 cubic yard

To use the table, figure the cubic feet of concrete needed for your project, then round the figure up to the nearest 10 cubic feet and divide by ten. Multiply this new figure by the numbers on the table to find how much of each ingredient you'll need. Round off any fractions to the next whole unit higher. If you order bulk materials by the cubic yard, remember that each cubic yard contains 27 cubic feet. Dry materials also are sold by the cubic-foot sack.

Building the forms

Following are instructions for building forms for a straight wall (no more than 3 feet high) that's to be poured on a cured footing (see pages 50–51). If you'd rather not build your own forms, you may be able to rent prefabricated ones—see "Concrete Construction Forms" in the Yellow Pages.

A straight wall form (see the illustration following) is made up of sheathing, studs, spreaders, wire ties, and (for thicker walls) wales. Sheathing forms the mold. Studs back up and support the sheathing. Spreaders set and maintain spacing and also prevent collapse of the form prior to the pour. Ties help hold the form together and resist the pressure of the wet concrete. Wales align the form and brace the studs.

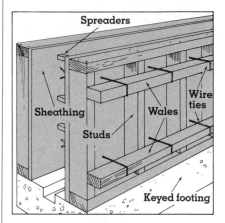

Forms for straight walls are built in paired sections—one for the front of the wall and one for the back. For each section, make a frame of 2 by 4s. The frame should be a workable length (8 feet is convenient) and as high as you'd like the finished wall to be. Next, nail in 2 by 4 studs; 16 to 19-inch centers are adequate for most do-it-yourself projects, but space closer if you plan an especially thick wall to be poured all at once.

The frame is now ready for sheathing; use either ½ or ¾-inch exterior plywood or 1-inch lumber. Lay the paired sections down, sheathing sides face to face, so that you can position the wales, if any, and drill holes for the ties. (Your local building department can tell you if wales are advisable for your project; they can also tell you how many you'll need and where to position them.)

Mark and drill a ⅛-inch tie hole in the sheathing along both sides of each stud, near the top and bottom; see illustrations on facing page. (If you're using wales, drill the holes as shown in the illustration above and toenail the wales in place.)

Make spreaders of 1 by 2s or 2 by 4s; cut them the same length as the finished thickness of your wall, and make enough so you can put them in about every 2 feet, horizontally and vertically.

Cut wire ties (8 or 9-gauge iron wire) long enough to encircle opposite studs (or opposite wales, if used), plus several inches to twist at the ends.

Tilt two sections upright on the footing, face to face, and spaced at wall thickness; tack the tops with several crosspieces to hold the sections in place. Thread a tie through the form and around the opposite studs or wales, and twist the wire ends together. Near the tie, wedge a spreader between the sections of the form and, if necessary, hold it in place.

Now put a stick between the two wires inside the form and twist them (as shown in the illustration following) until they are tight and the spreader is securely wedged between the form sections. Remove the

stick and repeat until all spreaders and ties are in place. You'll have to remove the spreaders as the pour is made, so tie long pull wires to any of them that will be out of reach, and hang the wires over the form top.

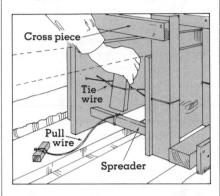

Add form sections to this initial one to make up the entire wall form; butt the sections and nail through the adjacent 2 by 4 studs. The running length of the side panels should be several inches greater than the finished wall so that you can cleat

in stop boards at the ends of your form (see the "Pour and tamp concrete" illustration below).

Just before pouring the concrete, coat the inside surfaces of the form with a commercial releasing agent or motor oil; this will make it easier to remove the form.

If you're building a low wall, your building department may not require the method described above. A simpler form, as shown below, may suffice.

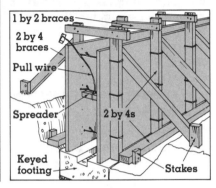

Steps in pouring a concrete wall

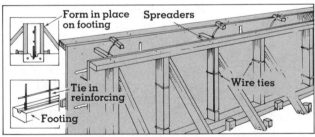

Mount forms on footing. Mount forms on footing and connect any required reinforcing bars to footing reinforcing as shown. Overlap reinforcing bars generously and tie at several places with wire.

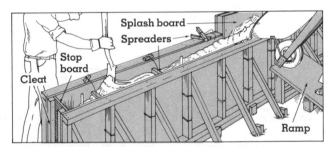

Pour and tamp concrete. Pour and tamp concrete in 6 to 8-inch layers, pulling spreaders out as you go and working concrete in and around any reinforcing bars. For smooth finish, work concrete well against forms.

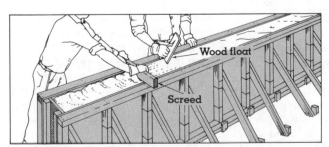

Screed, float, and trowel. Screed concrete level with tops of forms and smooth with float. Use trowel for extra-smooth finish. Cover and cure concrete for 4 days, using same method as for footings (see page 51).

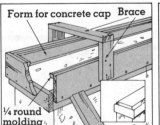

Two alternative caps. For a wood cap (right), set anchor bolts after wet concrete is screeded. Or cast an ornamental concrete cap: after wall forms are removed, build a form (left) on cured wall.

Walls: Brick

Building with brick is pleasant work—and relatively easy. The units are sized for easy one-hand lifting, and bricklaying takes on a certain rhythm once you get the hang of it. As with all building projects, planning and attention to detail are the keys.

Buying brick

There are two factors to consider in selecting brick for your wall: the type of brick and the size of brick.

Brick types. Because of the expense of shipping such a heavy material, brick is usually produced locally. No one type will be available all over the country, but there are two broad categories: common brick is most often used for general building; face brick is used where strict uniformity of appearance is required. Within each category, you'll find a wide variety of colors and textures.

Brick sizes. Most brick is made in modular sizes—that is, the length and width are simple divisions or multiples of each other. This simplifies planning, ordering, and fitting. The standard modular brick measures 8 inches long by 4 inches wide by 2⅔ inches high. Other modular sizes may be available at larger brickyards, in sizes ranging from 12 by 4 by 2 inches to 12 by 8 by 4. Note that all of these dimensions are nominal—they include the width of a standard ½-inch mortar joint, so the actual dimensions of the brick are reduced accordingly.

It is common for brick to vary somewhat from specified dimensions. To calculate the quantities of brick you'll need, consult your building supplier.

Bond patterns

Over the years, many brick bonding patterns have evolved: headers (bricks running perpendicular to the wall's length) and stretchers (bricks running parallel to the wall's length) have been used in a wide variety of combinations in each course, or layer, of bricks in a wall. In addition to the common American bond shown on the facing page, two others are often used for garden walls.

English bond alternates courses of stretchers and headers, forming very strong brick garden walls.

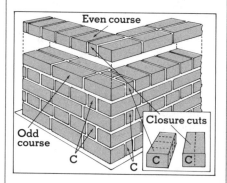

Flemish bond alternates headers and stretchers in each course. It is both decorative and structural.

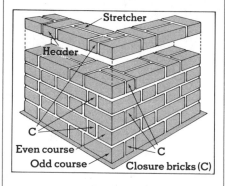

Mortar for setting bricks

Mortar is the bonding agent, the "glue" that holds masonry units together. Beyond this, it has several other functions: it seals out wind and water, compensates for variations in the size of masonry units, anchors metal ties and reinforcements, and provides various decorative effects,

depending on how the joints are finished (see page 57).

You can mix your mortar from bulk materials or purchase it dry by the bag. The most common mortar you can make from bulk materials is composed of cement, sand, lime, and water measured by volume in the following proportions:

1 part Portland cement
½–1¼ parts hydrated lime
6 parts clean, sharp-edged sand
Water as required

As the purpose of the lime is to make the mortar workable, you'll have to experiment with both the amount of lime and the amount of water you add. You want a mix that has a smooth, uniform, and granular consistency and—most important—that will stick to a vertical surface. An exact formula can't be given, because of such variables as weather conditions and the absorption rate of the bricks. Mix the dry ingredients thoroughly, using the minimum amount of lime, and add water (and more lime, if desired) a little at a time until you get the right consistency. Basically, mortar should be stiff enough not to drip, yet wet enough to stick. Mix small batches—about what you'll use in an hour.

To apply a mortar bed, scoop up some mortar with an 8 or 10-inch trowel. Turn the trowel over and spread the mortar, about ½ inch thick and 4 to 5 bricks long, with a zigzag motion of the trowel out to the edge. As you lay each brick, "butter" (apply mortar to) one end. See the facing page for more information.

Reinforcing brick walls

Walls more then 2 feet high need reinforcing. What's required may be as simple as a cap of header bricks (see facing page) or as exacting as engineered steel reinforcing. For higher freestanding walls, you have two choices: one is brick pilasters, or columns, built as part of the wall; the other is steel reinforcing. To make a choice and obtain instructions consistent with your local code, check with your building department or landscape architect.

Steps in building a brick wall

Poured concrete is recommended for your wall's footing (for instructions, see page 50). The footing needs to cure for about 2 days.

Distribute your bricks along the jobsite. Unless they're already damp, hose them down several hours before you begin work.

Mortar joints need to be "finished" periodically; see page 57 for instructions. If your wall is to have a corner, see page 56.

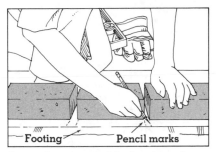

Lay a dry course. Mark outer edges of wall on footing with chalk line. Lay single "dry" course of bricks with ½-inch spaces for joints along full length of wall. Mark joints with pencil on footing.

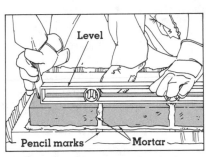

Lay first bricks. Take up bricks from dry course. Spread mortar on footing and lay first brick. Apply mortar to one end of second and third bricks, and set in place. Check that bricks are level.

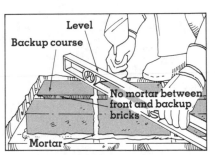

Begin backup course. Lay three backup bricks (there's no need to mortar the joints between courses). Make sure courses are level with one another, and check that overall width of wall is equal to length of one brick.

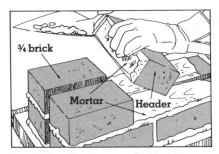

Begin header course. Cut two ¾ bricks (see page 40) to begin header course, and mortar in place. Lay four headers across width of wall, applying mortar as above. Finish joints as you go (see page 57).

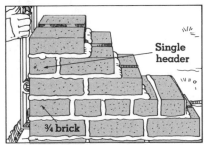

Complete lead. Continue laying stretchers until lead (end of wall) is five courses high, looking like steps. Note that fourth course begins with a single header. Now build lead at other end of wall.

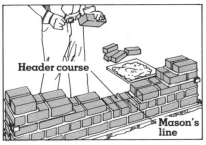

Fill in between leads. Stretch a mason's line between leads. Keep line ¹⁄₁₆ inch away from bricks and flush with top edges of course being laid. Lay bricks from ends toward center, course by course.

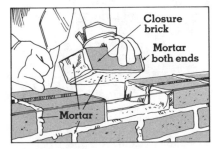

Close top course. Apply mortar to both ends of closure brick and insert it straight down. Repeat for backup course. To raise wall higher, repeat construction process, building new leads and filling in between them.

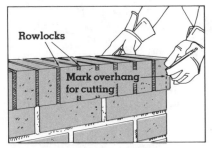

Plan cap. Lay "dry" course of header bricks on edge ("rowlocks") on top of wall, allowing for joints. If last brick overhangs end, mark it and cut off excess. Lay the cut brick three or four bricks from end.

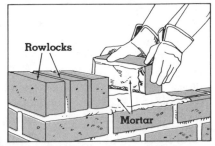

Set cap. Spread mortar on top of wall and begin laying cap. Apply mortar to face of each succeeding brick and set in place, squeezing some mortar out of joints. Finish joints as you go.

(Continued on page 56)

...*Continued from page 55*

Steps in building a brick corner

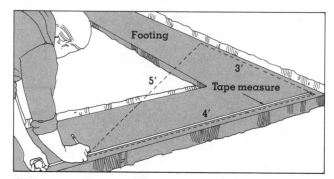

Checking for square. *After snapping chalk lines on footing for outer face of wall, measure 3 feet along one line and 4 feet along the other; distance between the two points should be exactly 5 feet.*

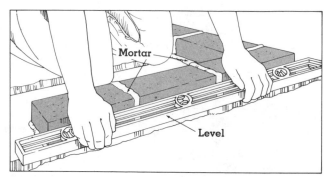

Start corner lead. *Lay single "dry" course of stretchers for entire wall and mark joints on footing (see page 55). Lay first brick at corner, then two bricks on each arm of wall. Make sure they're straight, plumb, and level.*

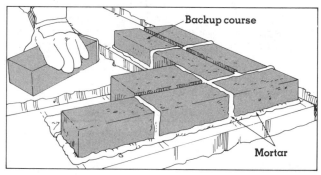

Lay backup course. *Spread mortar and lay backup course as shown, being careful not to disturb first bricks. There is no need to mortar the joints between courses. Check that all bricks are level.*

Start header course. *Cut two bricks into closures (¼ and ¾ pieces). Lay them as shown and complete lead header course. Now complete lead with three stretcher courses (see page 55), alternating position of corner brick.*

Special types of brick walls

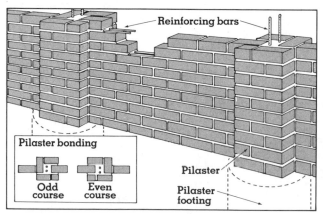

Panel walls. *In this type of brick wall, only the steel-reinforced pilasters have foundations. Support for the panels (which rest on the ground) comes from horizontal steel reinforcing, embedded in mortar joints, tying them right into pilasters. Panel walls can save much time and money, but sophisticated design and careful engineering are critical.*

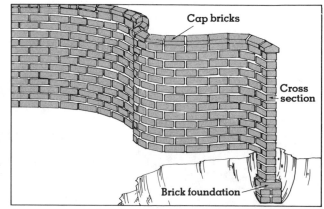

Serpentine walls. *The distinctive curve of this wall is actually an engineering feature; it helps wall to resist toppling and allows you to build a thin wall higher than you could otherwise. The wall shown is only 4 inches (1 brick) thick. The brick footing shown is easiest to use, but make sure that both footing and wall will satisfy local building codes.*

Steps in finishing mortar joints

Finishing—"striking" or "tooling"—mortar joints compacts the mortar and shapes it to shed water. Finishing is an important part of masonry because it contributes to the strength and weathertightness of the completed project. If you live in a freezing or rainy climate, use a concave or weathered joint, or a V-joint, as shown below.

Finishing tools vary, depending on the type of joint desired. Special tools are available, or you can use a trowel, a piece of wood, or a steel rod.

You can wait until the end to finish the joints, if your project is small. For larger jobs, though, you'll need to do this periodically as you work.

Mortar joints should be finished when the mortar is neither so soft that it smears the wall nor so hard that the tool shapes it with difficulty and leaves black marks. The mortar is ready to finish when it's "thumbprint hard"—when pressing on it leaves a slight indentation.

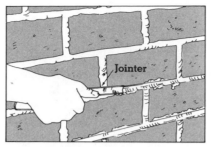

Finish bed joints. *When mortar is thumbprint hard, run jointing tool along bed (horizontal) joints first.*

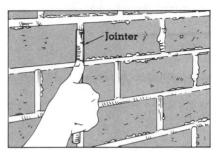

Finish head joints. *Now press and draw jointing tool along each head (vertical) joint.*

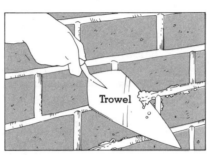

Cutting tags. *Slide trowel along wall to cut off tags (excess mortar). Then finish bed joints again.*

Brushing the wall. *Once mortar is well set, brush with a stiff brush or broom.*

Types of mortar joints

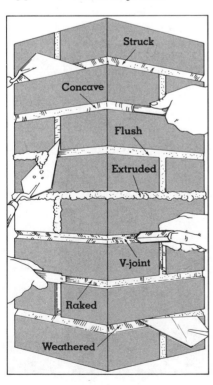

Struck joint provides dramatic shadow lines but tends to collect water. The joint is struck with the edge of the trowel; some compacting results.

Concave joint is made with a special jointer, as shown, or with a similar convex object. This joint readily sheds water and is well compacted.

Flush joint is made in the process of ordinary bricklaying; excess mortar is simply cut away with the trowel. This is not a strong joint because the mortar isn't compacted or waterproof.

Extruded joint results when mortar is allowed to squeeze out (extrude) from the joint as the brick is laid. It

has a rustic appearance but isn't waterproof, so it isn't suitable in rainy climates or where much freezing occurs.

V-joint is similar to the concave joint in both strength and weather resistance. Use a special tool (as shown), metal bar, or piece of wood to make this joint.

Raked joint casts dramatic shadows. Weather resistance is poor, though, and the joint is weaker than the concave joint and V-joint.

Weathered joint is struck from below with the trowel. It's the most waterproof joint of all and is somewhat compacted.

Walls:
Concrete block

The large size of concrete blocks makes for rapid progress in building. Most freestanding walls can be built with only one size of block—8 by 8 by 16 inches is common.

Alone, concrete blocks are comparatively weak because of their small bonding surfaces and thin shells. Where extra strength is needed, their hollow cores can be easily filled with steel reinforcing bars and mortar or concrete. The blocks become, in effect, permanent forms for what is essentially a poured concrete wall.

Blocks are available in an array of sizes and shapes. All except slump block have precise dimensions, making it easy to plan and design your wall. Because block sizes are standard, you'll have to maintain a consistent mortar joint (⅜ inch is common).

Types of blocks

When you visit your local masonry supplier, you'll find two broad categories of block: basic and decorative. Either type is suitable for most residential projects. Blocks vary with the locale, so all types may not be available in your area.

Basic block. In addition to the 8-inch-wide size, blocks come in 4, 6, and 12-inch widths. Standard blocks, available in stretcher and corner forms (see the illustration following), are molded with regular heavy aggregate and weigh about 45 pounds each. "Cinder" blocks are made with special lightweight aggregates and weigh considerably less. You may also find variations on the basic block, including bond-beam blocks (made with cut-away webs to receive reinforcing rods) and

cap blocks (to finish the top of the wall).

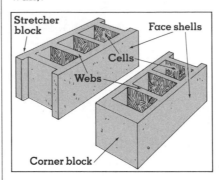

Stretcher block · Face shells · Cells · Webs · Corner block

Decorative block. Most manufacturers make a variety of decorative blocks with surface patterns that catch the play of light, enhancing a wall's appearance. The blocks shown below are examples of the many types available. Patterns vary with the locale.

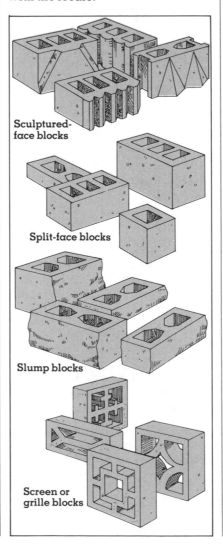

Sculptured-face blocks

Split-face blocks

Slump blocks

Screen or grille blocks

Mortar for concrete blocks

Use the same mortar you would use for brick (see page 54). Keep the mortar a little on the stiff side; otherwise the heavy blocks may squeeze it out of the joints.

Don't wet the blocks before laying them. The stiffer mortar and the lower rate of absorption of the blocks will keep them from soaking up too much water from the mortar.

Reinforcing concrete block walls

Block walls taller than 3 feet—and most block retaining walls—require reinforcing with steel (see page 63 for information on retaining walls). This is easily done by placing vertical bars in the footing while it is still soft so that they extend up through the block cores, which are then filled with mortar or concrete.

If horizontal reinforcing is required, use special bond-beam blocks with cut-away webs to allow the placement of horizontal reinforcing bars and concrete or mortar. You can also use special concrete block reinforcing for the horizontal joints of concrete block walls in place of bars (see illustration below). This type of reinforcing, made by welding lengths of heavy-gauge wire together, is available in long lengths from most concrete block suppliers. Building departments generally require it in every course of a screen-block wall.

Since code requirements vary and reinforcing is complex, be sure to check with your building department.

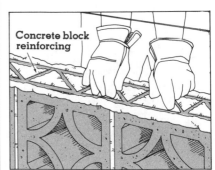

Concrete block reinforcing

How to build a concrete block wall

Building a concrete block wall is a lot like building a brick wall (see pages 54–57). Because your block work has to be precise, it's best to base the overall dimensions of your project on the block you will use. (If you have to cut blocks, use a power masonry saw.)

To build the recommended concrete footing for a concrete block wall, follow the directions on pages 50–51. Check on the positioning of the blocks by fitting a dry course on the cured footing; allow for ⅜-inch joints. Mark the position of each block and lay it aside. Then you can mix mortar, using the directions on page 54, and begin to build your wall. The process involves building up leads at either end of the wall, then filling in each course with the remaining blocks. For information on finishing joints, see page 57. Closure blocks are set in place using the same technique as for bricks (see page 55). A layer of ¼-inch metal screening underlies the top course of blocks. The screening holds the mortar that you pack into the top cells to give the wall a finished look.

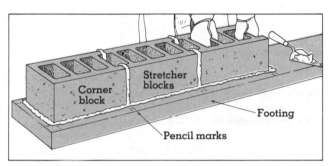

Spread a bed of mortar 2 inches thick and long enough for three blocks. Set blocks with mortared joints and tap firmly, making sure they're level and straight.

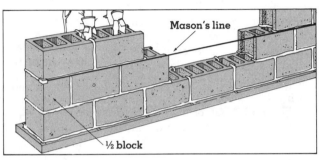

Build up leads at ends and corners with corner blocks first. Use half block to start each even course. Stretch a mason's line between corners to keep blocks straight.

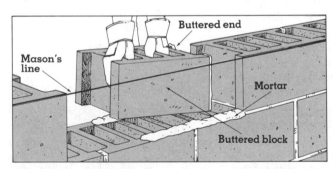

When setting each block, butter one end. Apply ribbons of mortar to blocks below. Place buttered end against end of last block set. Maintain ⅜-inch mortar joint.

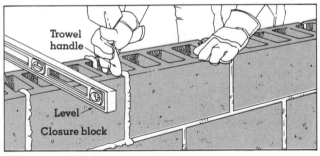

Butter both ends of closure block and set in place. Bed block by tapping it with trowel handle as you check it with level.

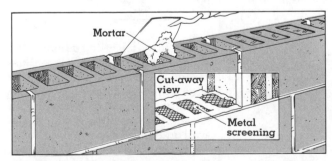

Position ¼-inch metal screening under mortar beneath top course. Fill cells of top course with mortar (supported by screen) and trowel smooth.

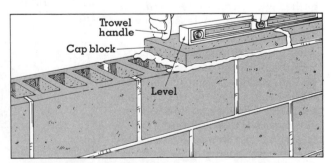

Spread ribbons of mortar along all top web surfaces of top course of blocks. Then cap top of wall with flat blocks for decorative finish.

Walls:
Stone

Because of the size, weight, and irregular shape of stones, building a dry or mortared stone wall is the most laborious and demanding kind of masonry work. But if you work carefully and methodically, you can build a stone wall that will grace your garden. You'll find examples of different kinds of stone walls in the color section (see pages 15, 16, and 18).

If you're considering a stone wall higher than 3 feet, or if you're faced with unstable soil conditions, consult a landscape architect or the building department.

Before you start building a stone wall, place your stones near the site for working convenience. Use the largest stones for the foundation course; this will not only spare your back but will also make the wall stronger. Reserve longer stones for bond stones—long stones running from the back to the front of the wall—and set aside broad, flat stones for the top.

Bonding in stone walls

As you build, be sure that vertical joints are staggered (the stones on each level should overlap the joints in the course below). Tie the wall together with bond stones, using as many as possible—at least one every 5 to 10 square feet.

To help secure the wall, "batter" (slope) both faces inward. The amount of slope needed depends on the size and purpose of the wall, the shape of the stones, and whether or not the wall is mortared. A good rule is 1 inch of batter for every 2 feet of rise, but this could vary with the shape of the stone—check with your landscape architect, building department, or stone supplier.

Make a batter gauge (see the illustration below) to check your work as you go; keep the outer edge plumb with a level.

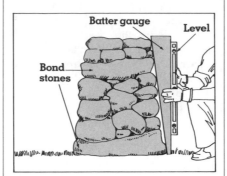

How to build a dry stone wall

In a dry stone wall, the stones hold each other in place by weight and friction. Though construction is simple, you must build with care to ensure a long-standing wall.

Unless your stones are very flat and rectangular, plan on a battered wall no higher than it is thick at the base. Very round stones require so much batter that the resulting wall may turn out to be no higher than one-third of its thickness.

Even in severe frost areas, dry walls are built on shallow foundations. Since the wall is flexible, the few stones dislodged by frost heave can be put back in the spring.

To build the wall, begin by laying the foundation stones in a shallow (about 6 inches deep) trench to help stabilize the wall. Begin with a bond stone at the end, then start the two face courses. Fill in the space between the face courses with tightly packed rubble (broken pieces of stone).

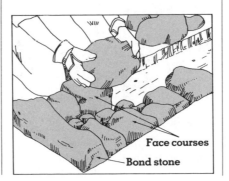

Build the second course by laying stones on top of the first course, being sure the vertical joints don't line up. Tilt the stones of each face inward toward each other. Use your batter gauge and level on faces and ends of the wall to maintain proper slope. Again, pack the center with rubble.

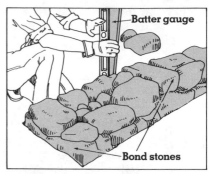

Continue in the same manner, maintaining the inward slope so gravity will help hold the wall together. Place bond stones every 5 to 10 square feet to tie the sides together. Use small stones to fill any gaps (see drawing below); tap them in with a soft-headed hammer to make the wall stronger, but don't overdo it—driving them in too far will weaken the structure.

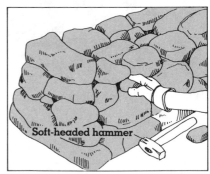

Finish the top, using as many flat, broad stones as possible (see illustration following). If you live in an area with severe freezing, consider mortaring the cap as shown in the inset. This will drain water off the wall and help prevent ice from forming between the stones. (Ice can push the stones apart and damage the wall.) For information on mortaring stones, see the facing page.

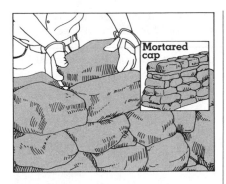

Mortared cap

How to build a mortared stone wall

Virtually any kind of stone can be mortared to make a stable wall. Bear in mind that the rounder the stones, the more mortar you'll need—the increase is dramatic. Many municipalities require an engineer's design and a building permit for walls over 3 feet high, so be sure to check. (The following instructions cover walls 3 feet high or less.)

You'll need a strong footing to support a mortared stone wall. You can build the footing by pouring concrete into wood forms (for instructions, see pages 50–51) or by filling the footing trench with concrete and stones. The more stones you use, the less concrete you'll need. Make sure the concrete fills the voids between stones. Level off the top of the footing to make a flat surface on which to build the wall.

For a 3-foot-high wall, the footing trench should be at least 12 inches deep (in freezing locales, the bottom of the trench must be below the frostline). Allow the footing to extend 6 inches beyond the wall's edges on all sides. Check with your building department for steel reinforcing requirements.

Mortar for stonework. The mortar formula for stonework is richer than that used for brick: use 1 part cement to 3 or 4 parts sand. You can add ½ part fireclay for workability, but don't add lime (or use mortar cement, which contains lime) because it might stain the stones. Keep the mortar somewhat stiffer than for brick.

Depending on the shape of the stone, a wall may be as much as one-third mortar because of the joints and voids. To plan for this, lay up a small section of the wall, note the amount of mortar used, and use this as a guide for the rest. It's not a bad idea to use this method even if you're working with well-trimmed stone. Every stone wall is a special case; laying a sample section will be your best guide to the amount of mortar you'll need.

Building the wall. Before you start building the wall, make sure your stone is conveniently nearby and that it's clean and dry (dirt and moisture interfere with the bond).

Spread a 1-inch-thick bed of mortar at one end of the footing and set the first bond stone, making sure it is well bedded. Now lay the stones in the first course of the front and back faces, spreading the mortar bed as you work. If you have to trim a stone to fit properly, chip away the excess material with a stonemason's hammer. (Be sure to wear safety glasses.)

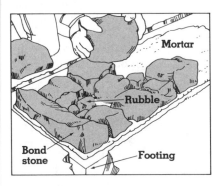

Mortar

Rubble

Bond stone

Footing

Pack the joints between stones with mortar, and fill the space between the front and back faces solidly with rubble and mortar.

For each subsequent course, build up a mortar bed over the previous course of stones and set the new course of stones in place just as you did for the first course. Remember to place a bond stone every 5 to 10 square feet and to offset the vertical joints. Work slowly, dry-fitting stones before you spread the mortar. You can save mortar by filling large joints with small stones and chips. As you work, check alignment and plumb or batter.

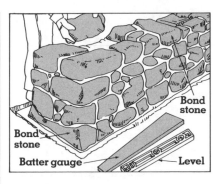

Bond stone

Bond stone

Batter gauge

Level

Very large stones may squeeze out the mortar in their joints. Preserve joint spacing by supporting such stones on wood wedges. When the mortar is stiff, pull out the wedges and pack the holes with mortar.

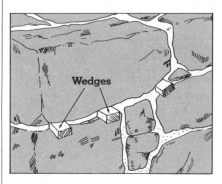

Wedges

After you've laid a section of wall, and before the mortar sets, use a piece of wood to tool the joints to a depth of ½ to ¾ inch.

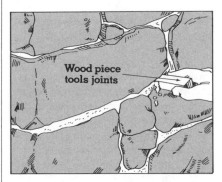

Wood piece tools joints

Wipe spilled mortar from the faces of the stones as you work. After finishing the joints, use a broom or brush to remove crumbs of mortar. Once the mortar has dried, wash the wall with clear water. If this doesn't remove mortar stains, use soapy water followed by a clear rinse. Avoid using muriatic acid and steel brushes.

Walls:
Adobe block

Adobe block walls carry the signature of the old Southwest. Today's adobe blocks are stabilized with an asphalt emulsion that makes them impervious to water—an advancement that makes building with adobe blocks a more practical choice and has led to their use outside the Southwest.

Sizes of adobe blocks

Adobe blocks come in several sizes, usually 4 inches thick, 16 inches long, and 4½, 5½, 7½, or 12 inches wide. If your building supply yard doesn't carry adobe blocks, see about having them ordered—or look into ordering them yourself from a manufacturer. In either case, check on transportation cost; it can be considerable. If you're interested in making your own adobe blocks, see page 46.

Mortar for adobe blocks

Traditional adobe walls were mortared with a crude mixture of water and the same soil used for the blocks. Today, we use a much more stable mortar (different from that used for adobe paving), containing the same soil used for the blocks, and mixed with cement, sand, and asphalt emulsion.

Follow these steps to make mortar for an adobe block wall: Mix 1 part cement, 2 parts soil, and 3 parts sand. Add 1½ gallons of asphalt emulsion for every bag of cement. Add enough water, a little at a time, to make the mortar spreadable but not runny. (The wetter the sand, the less water you'll need.) The soil gives the mortar an adobe color.

Ideally, the soil in the mortar should be the same as that used to make the blocks. The manufacturer of the blocks can supply you with a suitable soil and the asphalt emulsion. Don't use clay soil—particularly the kind commonly called "adobe." Many clay soils are detrimental to the mortar and blocks.

Reinforcing adobe block walls

You'll need to add strength to your adobe block wall with steel reinforcing bars laid in the bed joints (see the following section). Some walls may even require vertical reinforcing bars placed in the footing and extending up through holes drilled in the blocks. Ask your landscape architect or building department for exact reinforcing requirements.

Laying adobe blocks

Building an adobe block wall is similar to building a brick wall (see pages 54–57). The following instructions are suitable for walls up to 3 feet high. Consult a landscape architect or your building department if you want a higher wall, and—for either a low or high wall—check for reinforcing requirements.

The bond illustrated at right, called running bond, is an easy pattern to lay and is almost always used in adobe walls. You'll probably have to cut a few blocks to use at ends or corners. To cut adobe blocks, use the method for cutting bricks (see page 40) or use a masonry saw, available from a tool rental store.

Build a concrete footing for your adobe block wall following the directions on pages 50–51. Make sure the footing is on solid ground; adobe blocks are less resistant than bricks to flexing caused by soil movement. To keep the footing dry and extend the life of your wall, build the footing so that its top is 6 inches above the ground and it projects at least 6 inches behind and in front of the wall as well as at the ends.

When the footing is dry, build up

the leads at the ends and corners as you would for brick (see pages 55–56). Use ½-inch mortar joints. Then stretch a mason's line between the leads (see the illustration below) to guide you. Finish the joints as you would for brick (see page 57).

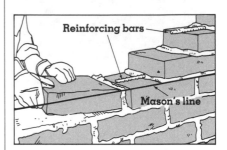

Reinforcing bars

Mason's line

Following the design specifications of your landscape architect or building department, build reinforcing bars into the leads and along the bed joints, as illustrated above. Overlap the ends of the bars and splice them with wire. For corners, prebend a length of bar and splice it to the ends of the bars in the bed joints (see the illustration below). The standard practice for splicing calls for an overlap 40 times the bar diameter—10 inches for ¼-inch bars. Wrap the wire around the bars to tie them together.

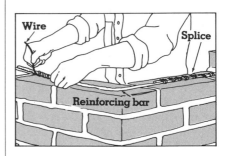

Wire

Splice

Reinforcing bar

Finishing adobe

Postpone any decision about finishing your adobe wall until you see how the completed wall looks in its setting. Plaster covered with whitewash is the traditional finish for adobe walls, but you can leave the finish off if it doesn't suit your garden. Once plastered or painted, a wall loses its rustic appearance, and the finish cannot be removed without ruining the texture of the block.

Retaining walls

If your home sits on a sloping lot or a hillside, a retaining wall may be part of your landscaping plans.

Simple wood or masonry (concrete, concrete block, brick, or stone) retaining walls, less than 3 feet high and on a gentle slope with stable soil, can be built by an experienced do-it-yourselfer. Remember that most communities require a building permit for a retaining wall and may require a soil analysis in any area suspected of being unstable. A wall in an unstable area and walls over 3 feet high must be designed and often supervised by a licensed engineer. In any case, professional advice for any retaining wall can save you time, money, and possible disappointment. So check with your building department or a landscape architect before starting a retaining wall project.

If a retaining wall project is intended to hold up a slope, extensive cutting and filling may be needed.

The hill can be held with a series of low retaining walls which form terraces, or with a single high wall (see the illustration below). Though the first is less risky, both disturb the hill and should be designed by a professional.

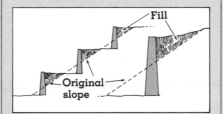

If space permits, the safest approach for the homeowner considering a retaining wall is not to disturb the slope at all. Build the wall on the level ground near the foot of the slope and fill in behind it (see the illustration below).

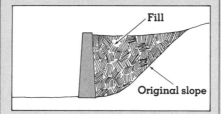

There are three basic types of retaining walls. Your choice will depend on your situation and budget.

The simplest and usually the cheapest wall relies on planks or timbers supported by posts to hold back the earth (see the "Wood retaining wall" illustration below). Since the weight of the earth behind the wall is transferred through the planks or timbers to the posts, they must be properly sized and set deeply and firmly into the ground. You can use this type of wall for leveling gently sloping areas, where you do not have to cut deeply into the slope.

Retaining walls intended to hold more radical slopes are either of the *mass* or *cantilevered* type. The mass type (see the illustration below) relies on its own weight to prevent it from tipping or sliding. To prevent water from accumulating behind the mass retaining wall (where the weight of the waterlogged earth might cause the wall to collapse), gravel and lengths of perforated drain pipe are placed behind the wall as illustrated. The pipes slope down from the center of the wall toward each end.

The cantilevered wall uses the weight of the earth, resting on a wide footing behind it, to hold the wall in place. To provide for drainage, the wall is pierced with weep holes to allow water to escape (see the illustration below). Like the mass wall, the cantilevered wall also has a drainage bed of gravel behind it.

Three types of retaining walls

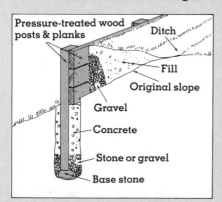

Wood retaining wall. *Use for mild slopes where ground is level behind wall top. Resembles a heavy fence, with posts sunk half their length into the ground, and closely spaced.*

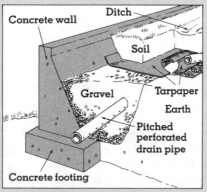

Mass retaining wall. *Earth presses down on pyramidal wall and wide footing to hold wall in place. Use broken concrete pieces and stones in the poured concrete to lower cost.*

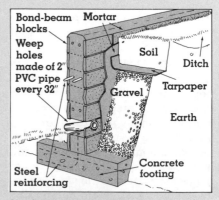

Cantilevered retaining wall. *Built of reinforced masonry (concrete block is shown), this wall stays in place because of weight of earth on wide footing.*

FENCES & GATES

Simple, effective & easy-to-follow building techniques

Before you set a post or pound a nail, you need to choose a fence or gate style and building materials. First look into local codes and ordinances that may influence those decisions, and learn how height, material, setback, and other requirements affect your project. Then you'll be ready to tackle the building stages: plotting the fence, installing fence posts, and adding rails and siding.

This chapter explains and illustrates how to build a straight fence on level ground, and gives tips on how to build and repair gates.

Materials. For a long-lived, weatherproof fence, use either pressure-treated wood (which has had preservative forced deeply into it) or redwood, western cedar, or some kinds of cypress (specify heartwood; it is decay and termite resistant).

Use only rust-resistant hardware for your fence—hot-dipped galvanized or aluminum nails and galvanized bolts, screws, hinges, and catches.

Fences & gates: Plotting your fence

The first step in building your fence is to determine the exact course it will take and mark that course with stakes and mason's line.

If you're building a new fence (or replacing an old one) on or next to a boundary or property line and the original survey markers can't be found, have a surveyor locate and mark the boundary. Unless you and your neighbor are making a joint project of a boundary fence, it's wise to eliminate possible future disagreements by building the fence 1 to 2 inches on your side of the line.

Plotting a straight fence. Mark each end or corner post location with a solidly driven stake. Run mason's line between stakes, drawing it tight and tying it firmly to stakes.

Locate sites for remaining posts by measuring along the line and marking post centers on the mason's line with chalk. Using a level or plumb bob, transfer each mark to the ground and drive in a stake to mark the post location.

Plotting a right angle corner. If your layout calls for corners that form exact 90° angles, use the 3-4-5 rule to square the corners (use 6-8-10 or 9-12-15 for greater accuracy).

Establish the first fence line (point A to point B), following the directions for "Plotting a straight fence." Locate the second fence line (point B to point C) approximately perpendicular to the first. Build a batterboard beyond point C and tie the line from stake B to it; remove stake C.

On the first fence line, drive in a stake 4 feet from stake B (point D). Drive a nail into the top of each stake so that the nail in stake D is exactly 4 feet from the nail in stake B.

Hook the end of one steel tape measure over the nail in stake B, and hook the end of another tape over the nail in stake D. Have a helper pull out both tapes until the 3-foot mark on the tape hooked to point B coincides with the 5-foot mark on the tape hooked to point D. Transfer this point to the ground using a stake with a nail at the top (point E). Adjust the end of the line attached to the batterboard until it passes directly over the nail at point E. The two fence lines now meet at a 90° angle.

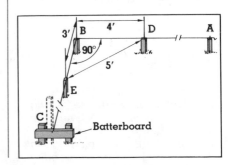

Twelve fence styles

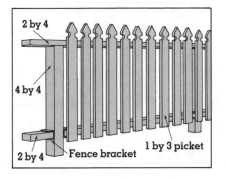

Picket

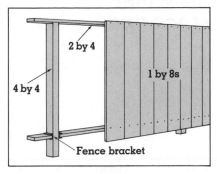

Post & board

Solid board

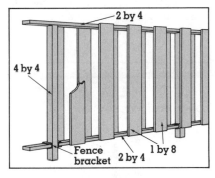

Alternate board

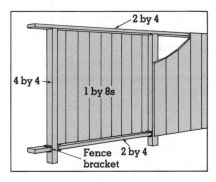

Alternate panel

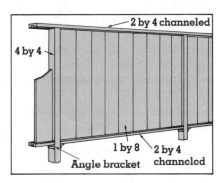

Good neighbor (same both sides)

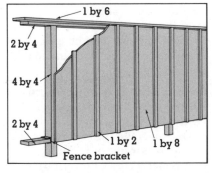

Vertical board & batten

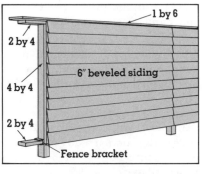

Bevel siding

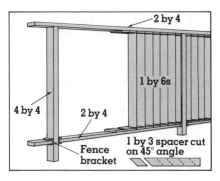

Louver fence

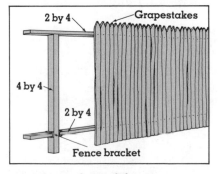

Grapestake

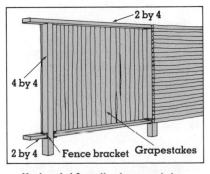

Horizontal & vertical grapestakes

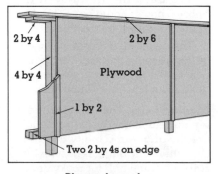

Plywood panel

Fences & gates: Installing fence posts

The most important part of fence building is setting and aligning fence posts. Posts that aren't set firmly in the ground or aren't set deep enough may allow an otherwise solidly built fence to tilt or collapse. And to prevent problems when you install rails and siding, be sure posts are plumb (aligned vertically in their holes) and located exactly on line.

Digging postholes

For fences less than 3 feet high, sink your line posts at least 18 inches into the ground. For most fences between 3 and 6 feet, set line posts at least 2 feet deep. End posts and gate posts need to be stronger and should be set 12 inches deeper than the line posts.

Fences taller than 6 feet or fences subject to unusual stress (strong winds, unstable soil, or heavy siding) should have one-third of their post length in the ground and set in concrete.

To prevent fence posts from settling, dig each hole deep enough so you can set the bottom of the post on a large, flat-topped stone or on 6 inches of well-tamped gravel. If you're building in an area that freezes, posts should be sunk at least 12 inches below the frost line.

The diameter of a posthole should be 2½ to 3 times the width of a square post or the diameter of round posts (10 to 12 inches for a 4-inch post).

Digging tools. The two most popular hand tools for digging postholes are the auger and the two-handled clamshell. The auger is best for loose soil, the clamshell for rocky or hard soil. If soil is too rocky or hard to dig, use a digging bar or jackhammer to break it up.

If you have more than six holes to dig and the earth isn't too rocky, let a power digger save you time and effort. Models that can be operated by one or two persons are usually available at tool rental shops; so are jackhammers and hand tools.

Setting posts

For strongest installation, set fence posts in concrete. If the soil is stable (not subject to frost heaving, sliding, or water saturation), you can set posts for lightweight or low fences (less than 3 feet high) in earth-and-gravel fill. The concrete or earth-and-gravel fill should be angled down from the post to ground level to divert water away from the post. Though you can set posts with their tops at the exact height wanted, you'll find it easier to allow a few extra inches in height and then cut them to the correct height later.

Setting posts in concrete. If this method is best for you, plan to work carefully to prevent concrete from creeping under the posts, where it can retain moisture and speed decay.

To mix your own concrete, use 1 part (by volume) cement, 3 parts sand, and 5 parts gravel. Keep the mix rather dry so earth from the side of the hole will not mix in and weaken the concrete. To save time and effort, you can make the concrete using bagged ready-mix. You'll need about a bag of mix for a 4-inch post set 2 feet deep.

Place a large and relatively flat-topped stone in the bottom of the hole. Fill in with gravel (about 6 inches) until it's level with the top of the stone. Set the post on the stone and shovel in 4 more inches of gravel, tamping it well with a 2 by 4. Plumb and align the post, using one of the methods described under "Aligning the posts," facing page. Finish filling the hole with concrete—2 to 3 inches at a time, tamping it in. Add another 1 to 2 inches of concrete above ground, sloping it as shown, for water runoff.

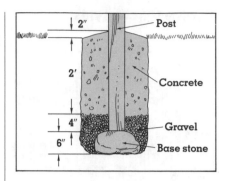

If the post hole ended up a little larger than you planned for and you find you're running short of concrete, don't rush back to the building supply store. Just add enough washed rocks to extend the concrete (see illustration below).

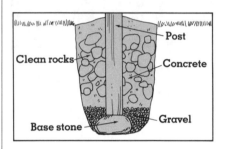

Be sure to double check your post for plumb and for alignment. You can force the post into a new position for about 20 minutes after pouring. Wait at least 2 days before installing rails and siding.

Setting posts in earth and gravel. Set the base stone and post in the hole, shovel in about 6 inches of gravel, and plumb and align the post (see directions on facing page). Dampen the gravel and tamp it with a length of 2 by 4. Continue filling with a mixture of earth and gravel, dampening and tamping firmly

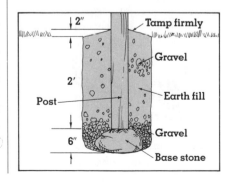

every 2 to 3 inches. After the hole is filled and the top sloped, check plumb and alignment again.

To minimize side movement of posts set in earth and gravel, you can add rocks to the fill as you near ground level. Fill the voids with earth and tamp it well. If your soil is light and sandy, you might consider nailing a 1 by 4 cleat on each side of the post before filling the hole (see illustration). The cleats should be long enough to fit snugly into the hole.

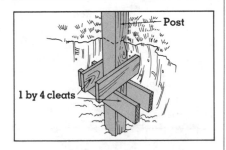

1 by 4 cleats
Post

Preventing frost damage. In freezing climates, frost heave can be a major problem. When water in the ground freezes and expands, it can push a fence post right out of its hole. And water that freezes and expands inside of wood posts can crack the concrete around them.

To minimize damage from frost heave, you can use 16-penny common nails to stud the sides of the post that will be below the frost line. Before doing this, though, make sure the post hole extends 12 inches below the frost line (your building department can help you determine this line). The nails will embed in the lower concrete collar, increasing resistance to frost heave.

To prevent concrete from cracking when wet posts freeze, you can cut shingles to the post width, coat them with motor oil, and temporarily nail them around the post before you set the post. When the concrete has set, pull out the shingles and seal the open spaces with caulking or tar.

(Both of the methods described above are shown in the illustration at the top of the next column.)

Set the readied post in gravel, as shown in the illustration following. Then plumb and align the post, us-

ing one of the methods described on this page. Use concrete to fill the hole up to the frost line, tamping well so there are no voids. Next, use earth to fill the hole to within 6 inches of the surface, adding 2 to 3 inches at a time; tamp well as you add each layer. Fill the rest of the hole with concrete, sloping the top so water will run away from the post. Double check the post for plumb and alignment before the concrete sets.

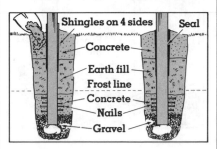

Shingles on 4 sides
Seal
Concrete
Earth fill
Frost line
Concrete
Nails
Gravel

Aligning the posts

Fence posts must be aligned so they're exactly vertical and in a straight line. It's much easier to do this if you have help, though you can do the job yourself if necessary. Use one of the following methods to align your posts.

Corner or end post method. Begin with two corner or end posts. Position them with their faces in flat alignment and plumb them with a level.

Next, nail a 2-inch-long 1 by 2 spacer block to each post—about 1 foot above the ground (see illustration below). Stretch and tie a mason's line between the posts, making sure the blocks are between the line and the posts.

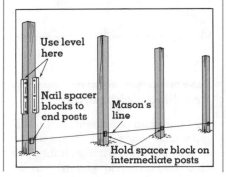

Use level here
Nail spacer blocks to end posts
Mason's line
Hold spacer block on intermediate posts

Set and align the intermediate posts so their faces are exactly the thickness of a 1 by 2 block (¾ inch) away from the lines. Before and after filling the hole, check each post for plumb and alignment, using a level on two adjacent faces.

Offset method. For long sections of fence (100 feet or more), the corner post method may be impractical because the mason's line will sag. One way to prevent this is to use the corner post method but build the fence in sections shorter than 100 feet.

Or you can set and align the posts successively, starting at one end of the fence line. All you do is move the stakes used for plotting the fence half the thickness of the posts to one side of the post centers. Then reconnect the mason's line, and align and set the posts, one by one, using the line as a guide. Be careful that you do not move the line out of alignment as you set the posts. To avoid this, use spacer blocks as described in the corner post method (preceding). Be sure to check each post for plumb, using a level on two adjacent faces.

Bracing method. You can set fence posts by yourself if you align your posts according to one of the two preceding methods and then use braces to hold them in position while you fill the holes.

Drive stakes into the ground in line with two adjacent sides of the post (see illustration below). Nail the end of a 6-foot 1 by 4 to each stake with a single nail so that the braces can be moved into position on the post. Plumb and align the post; then nail the other ends of the braces to it to hold the post in position while you fill the hole.

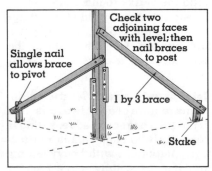

Check two adjoining faces with level; then nail braces to post
Single nail allows brace to pivot
1 by 3 brace
Stake

Fences & gates: Attaching rails & siding

Once the fence posts are set and aligned (see pages 66 and 67) the most difficult part of fence building is over. Next you'll attach the rails, install kickboards (if used), and secure the siding (boards, pickets, panels or other material).

Attaching the rails

Be careful when you fit, level, and fasten the rails. This part of the framework ties the fence together and must support the weight of the siding.

Adjusting posts for height. If you didn't set the tops of your posts at the finished height, now is the time to cut them. Use a combination square to mark the height you want on all four sides of an end post. Tie a mason's line to the post at your mark and stretch the line to the other end post if it's less than 35 feet away. If it's more than 35 feet away, use an intermediate post.

Hang a line level in the middle of the line. Have a helper pull the line tight and move the end of the line up or down the post until the line is level. Mark this point on the post, tie the line to it, and mark the intervening posts where the line touches them. Using a combination square, extend the mark around all four sides of each post as a guide for cutting. Continue this process until all posts are marked. Cut the posts at the marks, using a crosscut or power saw.

Joining rails to posts. Most fence designs require at least two rails, and if the rails are set on edge, siding is less likely to sag. This is particularly important with heavy siding, and if it's especially heavy, you may

need to add a third rail or install a wood or masonry support under the center of the bottom rail.

Butted rails should be cut to fit snugly between posts. Attach them to the posts with fence brackets or angle brackets, or by toenailing them. If you toenail the rails, nail 2 by 4 cleats to the posts first to help support the rails. Make sure rails are level before fastening them.

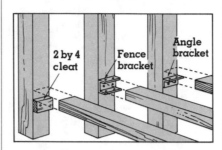

Lapped rails are the easiest to install and are most often used with vertical board fences. For a stronger fence, be sure the rails are long enough to span at least three posts, and stagger the joints of top and bottom rails.

Level the rails and nail them to the posts with hot-dipped galvanized common or box nails that are at least three times as long as the thickness of the rails. If rails lapped on tops of posts meet at a corner, miter the ends. If rails meet on an intermediate post, butt the ends at the middle of the post.

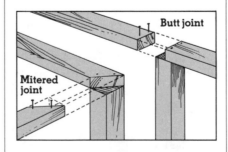

Installing kickboards

On board and panel fences, the siding usually ends 6 to 8 inches above the ground to keep it from rotting. To close this gap, a kickboard may be

nailed to the bottom fence rail before the siding is attached (see illustration below).

Kickboards are typically 1 by 8s or 1 by 10s of decay-resistant wood. They can be either centered under the bottom rail and secured with a cleat, or nailed to the faces of the posts and the bottom rail. Allow the kickboard to extend 4 to 6 inches into the ground to discourage animals from digging under the fence.

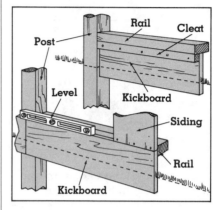

Attaching siding

This is the easiest part of building your fence, but don't be tempted to start nailing before you check to see that the framework is just right. If it isn't, the siding will make any problem all too apparent.

Attaching boards, slats, pickets. To keep boards, slats, or pickets in vertical alignment while nailing them to the rails, use a level to check each board for plumb before nailing it. Make sure all the boards are the same length. Then stretch a mason's line tightly along the fence where you want the bottom of the boards to end, checking the line with a level. Align the bottoms of the boards with the line or, if you're using kickboards nailed to the face of the fence, make sure their tops are level, and then rest the siding on top of them.

Attaching panels. If you're using panels for siding, you'll need a helper or two to lift the panels into position and hold them while you level and nail them in place.

Building a gate

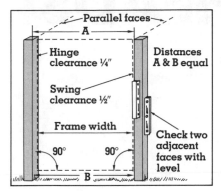

Set and align gate posts *vertically, with inside faces parallel and with distances A and B equal to gate width plus hinge and swing clearances.*

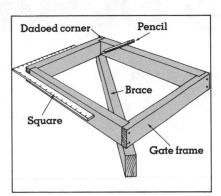

Build gate frame, *making sure corners are square. Mark, cut, and attach diagonal brace from top of latch side to bottom of hinge side.*

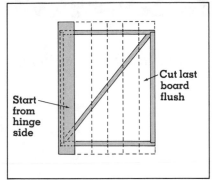

Attach gate frame *to post with at least two heavy-duty hinges. Nail siding to frame, starting at hinge side; cut last board flush.*

Repairing a gate

If you have a gate that binds, won't latch, drags on the ground, or has sagged out of shape, consider these ways to repair it.

Check the hinges—either replace them with heavier hinges or replace the screws with longer ones or with bolts. You can strengthen two weak hinges by adding a third hinge of the same type slightly above the gate's midpoint. Any gate more than 5 feet high or 3 feet wide needs three hinges.

If a gate post is leaning, you can dig down to its base, straighten the post, and fill the hole with concrete (for specific instructions, see page 66). Or if one post has simply leaned toward the other gate post, you can straighten it and hold it upright with a turnbuckle and a steel wire or rod running from the top of the leaning post to the bottom of the second or third post down the line (see illustration at right, above).

Most wooden gates aren't overly strong and, with weather and hard use, will eventually sag out of shape. You can square up a gate and prevent it from sagging again by marking and cutting a 2 by 4 brace to fit between the top of the latch side of the gate and the bottom of the hinge side (see illustration, bottom left). Use waterproof glue and screws or nails to hold the brace in place.

Another way to stop a gate from sagging is to use a wire and turnbuckle assembly, running it from the bottom of the latch side to the top of the hinge side (see illustration, bottom right).

If your gate works well in dry weather but won't close during the rainy season, plane or cut off the sides so that in dry weather you have a clearance of ¼ inch on the hinge side and ½ inch on the latch side.

Conversely, a fence may shrink so much in hot dry weather that the gate's latch doesn't catch. In this case, relocate the latch or replace it with a longer one.

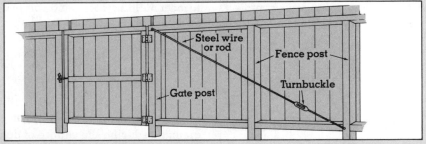

Use steel wire *(or rod) and turnbuckle to hold gate post upright; attach one end of wire support to top of leaning post, and other end to second or third fence post away.*

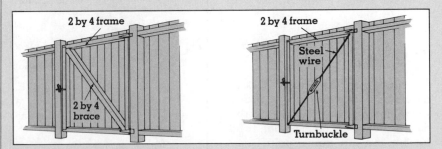

Correct a sagging gate *with 2 by 4 brace or with steel wire (or rod) and turnbuckle; attach each brace in directions as shown.*

DECKS & OVERHEADS

A practical guide to foundations and framing

With good planning, a few tools, and basic carpentry skills, you can easily build a wood deck or overhead. This chapter is a guide to building a wood deck attached to a house wall, or a wood overhead that's freestanding or attached to a house wall. Simply defined, an overhead is a structure built over your head—a patio roof, an arbor for your favorite vine, or a gazebo.

The emphasis in this chapter is on building, but design and planning affect construction. You may want to consult a landscape architect when planning your project, especially if your site is unstable or sloping.

Keep in mind that although you build a deck or overhead from the ground up, you must design it from the top down. First, choose the kind of decking or roofing you want (see pages 76 and 79). That decision will influence the sizes and spacings of the joists or rafters, beams, posts, footings, and piers.

As you plan your deck or overhead, check with your local building department to find out about regulations affecting the size, design, and construction of your project. In most communities, you'll have to meet building codes and obtain a building permit before you begin work.

It's best to use heart redwood or cedar, or pressure-treated lumber for exterior construction. Redwood and cedar are naturally resistant to decay and insect infestation; pressure-treated lumber is impregnated with chemicals to resist decay and insect infestation. To prevent the wood from becoming stained, use galvanized fasteners.

Read the entire chapter before you begin building; you'll find that two construction methods are discussed—one using joists and rafters, the other using posts and beams. The information on building a foundation and framing techniques applies to both methods.

Decks & overheads: Foundations

The foundation for a wood deck or overhead is a series of footings and piers designed to distribute the structure's weight on the ground and anchor the structure against settling, erosion, and wind lift. The foundation also isolates the wood framing from direct contact with the ground, reducing the chance of decay and insect infestation.

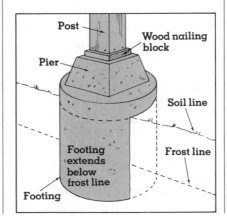

Building codes govern the size and spacing of footings and piers and specify how deep into the ground they must go. In general, footings must extend into solid ground or rock, and in cold climates, below the frost line so they're not disturbed by frost heave. Check with your building department for requirements in your area.

Locating the footings

From your plans, you'll know how many footings you need for your deck and how far apart you must place them. Work with care and patience to transfer these measurements to the ground.

First, attach a ledger to the side of the house (see page 73). With the ledger in place, square up the corners of the deck at each end, using the 3-4-5 rule:

Drive one nail into the end of the ledger to mark one corner of the deck, and another nail into the ledger exactly 3 feet away. Hook the end of the steel tape measure over each nail. Have a helper pull out the tapes until the 4-foot mark on the tape attached to the corner nail intersects the 5-foot mark on the other tape. Drive a stake into the ground at this point (see drawing below). This triangulation method works in any multiple of 3-4-5—for instance, 6-8-10 or 9-12-15. For maximum accuracy, use the largest ratio possible.

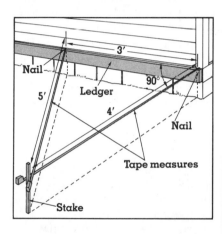

Measure the distance to the outside corner along the line that connects the end of the ledger and the stake, and drive in another stake to mark the corner. Then repeat the entire process at the other end of the ledger.

Build batterboards at each outside corner, using 2 by 4s for the stakes and 1 by 4s for the crosspieces (see illustration following). Locate the batterboards about 18 inches from each corner stake. Level the tops of the crosspieces with each other and with the top of the ledger, using mason's line and a line level.

Run mason's line from the nail at one end of the ledger to the opposite batterboard, making sure it passes directly over the corner stake (see il-

lustration below). Use the same method to mark the other side. Mark the outside edge of the deck with mason's line attached to opposite batterboards.

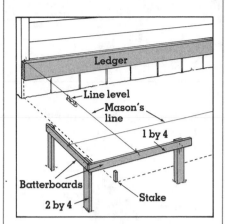

Measure the diagonal distance between opposite corners and adjust the lines until the distances are equal (see illustration below). Your deck outline is now square.

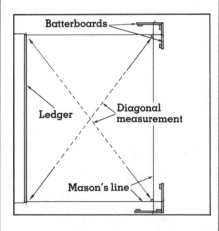

With a level or plumb bob, plumb down from the intersections of the mason's lines to recheck the corner stakes; these mark the locations of the corner footings. Measure along the lines and plumb down to locate any other perimeter footings required by your design.

Building the foundation

Most building codes require masonry foundations. Concrete is the most common material, primarily because it's the easiest to use.

If you want to mix your own concrete for footings and piers, use 1 part cement, 2 parts clean sand, and 3 parts gravel. Add water, a little at a time, as you mix. The concrete should be plastic, but not runny. You can also use dry-mix or transit-mix prepared with the same proportions of cement, sand, and gravel listed above. For more information about mixing and working with concrete, see pages 36 and 52.

There are three different methods for building a foundation: you can place ready-made piers on wet concrete footings; you can bond ready-made piers to dry concrete footings; or you can pour the footings and the piers at the same time. Ready-made piers are available at building supply outlets, and buying them is much more convenient than making your own. These piers usually have wood nailing blocks already embedded for toenailing wood framing posts.

Placing ready-made piers on wet footings. Begin by digging holes for the footings down to solid ground. Then fill the holes with fresh concrete to within about 6 inches of ground level.

Soak ready-made piers well with a hose before placing them. Position them 5 to 10 minutes after the footings have been poured. Then, check for level.

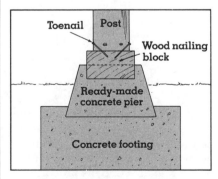

Bonding ready-made piers to dry footings. If you allow the footing to harden and add the piers at a later date, an additional step is required to bond ready-made piers to footings. Drench the top of each footing

Five kinds of post anchors

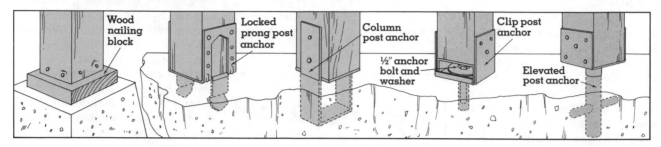

and the bottom of each pier with water. With a creamy paste of cement, coat both surfaces ½ to 1 inch thick; set the pier in place on the footing. Repeat with each footing and pier, each time checking to be sure the top of the pier is level.

Pouring footings and piers at the same time. If you decide to pour both piers and footings, place and level the piers' wood forms over the wet concrete footings, inserting any steel reinforcing that's required to strengthen the link between footings and piers (see illustration following). Then fill the forms with con-

crete and use a screed or straight board to level the wet concrete flush with the tops of the forms.

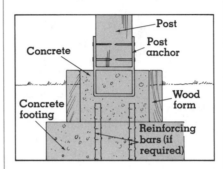

Immediately after filling and screeding, embed wood nailing

blocks or metal post anchors in the wet concrete piers (see "Five kinds of post anchors," above). Wood nailing blocks (adequate for the posts of a ground-hugging deck) are the simplest and least expensive to use. Metal post anchors are more suitable for raised decks, hillside decks, or high overheads. Use a carpenter's level to level the wood nailing blocks or post anchors in the concrete while it's still plastic.

To cure the concrete, leave the forms on the new piers and keep them damp for at least a week. For more information on curing concrete, see page 39.

Wood steps

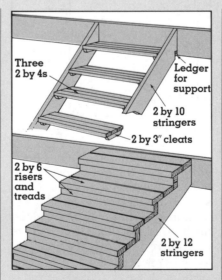

Following are instructions for building straight-run steps that lead from a deck to the ground. Because wood steps must be connected to the substructure, it's easier to build them before you nail decking. (For information on determining the number of steps needed and the proportions of the treads and risers, see "Ideas for garden steps" on pages 42–43.)

Two of the most common types of straight-run steps are shown in the following illustration. Steps with treads supported by cleats are much easier to build than those with treads cut into stringers.

Make stringers from 2 by 10s or 2 by 12s. Use galvanized bolts or metal joist hangers to secure tops of stringers to a deck beam or joist, as shown in the following illustration. (When bolts are used, the first tread is below the deck surface; with joist

hangers, the first tread must be level with the deck.) Anchor lower ends of steps to wood nailing blocks embedded in a concrete footing. If the steps are more than 4 feet wide, you'll need a third stringer in the middle.

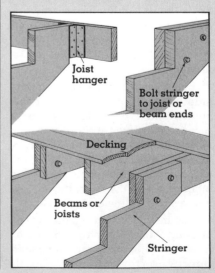

Decks & overheads: Framing

The substructure of a deck and the supporting framework of most overheads are similar. Ledgers, posts, and beams serve the same purpose for both decks and overheads, and the joists of a deck are equivalent to the rafters of an overhead. Post-and-beam construction (see page 75) can also be used for some decks and overheads, depending on the design; this type of construction doesn't require joists or rafters. The following instructions for framing apply generally to both decks and overheads. For additional information on framing, see pages 77 and 78.

Ledgers

If you want to attach a deck to your house, you must install a ledger—and you may have to install one for an attached overhead (see "Attaching ledgers for overheads," following). Ledgers—usually made of 2 by 4s or 2 by 6s—are fastened to the house framing through the exterior wall with lag screws, which must be long enough to go through the siding of the house and into the framing at least 1½ inches. To attach a ledger to a masonry wall, use expansion shields and lag bolts.

Attaching ledgers for decks. The most secure way to attach a deck to a house is to fasten the ledger to the floor framing or the wall studs. Wall studs are usually located on 16 or 24-inch centers. Floor framing members, often 2 by 10s, generally have about 1½ inches of subflooring and flooring on top of them. About 6 inches below the interior floor level you should find a framing member in which you can anchor your ledger.

A window sill makes a handy reference point for positioning the ledger (see illustration below). Brace the ledger in position and temporarily nail once at the ledger's approximate center. Level the ledger with a carpenter's level, and temporarily nail the ledger at both ends.

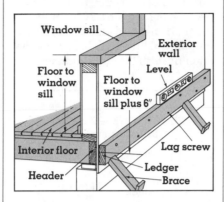

Drill pilot holes for the lag screws through the ledger and into the floor frame header of the house; drill clearance holes through the ledger. Lag screw the ledger in place and remove the braces.

Attaching ledgers for overheads. If your eave line is high enough for adequate headroom, you can easily attach a ledger for an overhead to wall studs—or, in a two-story house, to floor framing (to locate floor framing, see above). Measure first to make sure there'll be room for rafters between the ledger and eave line. If the eaves aren't high enough, one way to solve the problem is to mount a girder on the roof to support the rafters, using metal saddles, as shown below (you can have these made at a steel fabricating shop).

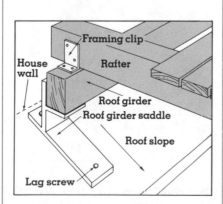

If your eave line is high enough to accommodate both ledger and rafters, brace, temporarily nail, and level the ledger as you would for a deck (see "Attaching ledgers for decks" on this page). Drill holes for lag screws, and lag screw the ledger to the wall, making sure you fasten to studs—otherwise the ledger is apt to give way. You can locate the studs (usually 16 or 24 inches on center) by observing the nail patterns in the siding or by measuring from corners, doors, and windows. Remove braces when the ledger is attached.

Attaching ledgers to masonry walls. The method for attaching a ledger to a masonry wall is similar to the methods just described. First, drill holes in the ledger at a minimum of 2-foot intervals. Brace and level the ledger in position and mark the location of the holes on the wall. Remove the ledger and drill holes for expansion shields, using a masonry bit; insert expansion shields in the holes. Brace the ledger while you lag screw it in place.

Flashing. Ledgers for decks fastened to a house wall must be covered with metal galvanized flashing to protect them from rain and snow. (Ledgers for overheads usually are protected by the eaves.) Form the flashing to fit, as shown below. Fit the flashing in place, caulk the top edge, and nail it in position with galvanized nails long enough to penetrate at least 1 inch into wall studs or other structural members. Cover the nail heads with caulking compound.

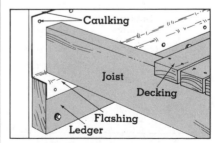

If the house is finished with shingles or lapped siding, you can use a simpler method for attaching flashing than the one described above. Just slip the top edge of the flashing up under the bottom edges of the shingles or siding as far as possible.

(Continued on page 74)

...Continued from page 73

Posts and beams

The most commonly used material for support posts is wood (though concrete and steel can also be used). For most decks and overheads, 4 by 4 posts are adequate.

Measuring for height. Accurate measuring of post height is an important phase in building a deck or overhead. Without it, your deck or overhead may be the wrong height, it may not drain properly, or it may be uneven.

Below are two methods for measuring. The illustrations show a deck (or overhead) attached to a wall. If your project is freestanding, you'll need to mark the desired height on one post and then use a mason's line and a line level to mark the other posts at the same height.

Method 1: Set a post in place, check for plumb, and brace it temporarily. Run a mason's line from the top of the ledger to the post and level it with a line level. Establish a mark on the post that is even with the top of the ledger. From that mark, subtract the thickness of the support beam you will use (see illustration below) and make a new mark. Set the other posts temporarily in place, and use a mason's line and a line level to mark the other posts at the same height. Take down the posts and cut off the excess.

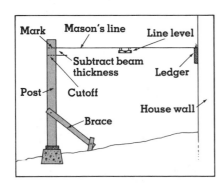

Method 2. Rather than setting up the post temporarily, you can determine post height by this procedure: After the ledger is in place, measure down from the top of the ledger to the ground ("X" in the illustration following). From this point, run a level line out to a footing and measure how far the top of the footing is below this point ("Z" in the illustration). To this distance, add the distance between the ground and the top of the ledger, and subtract the thickness of the beam; this would be 7¼ inches for a 4 by 8 ("Y" in the illustration). Cut the post to this length. Calculate the lengths of the other posts in the same manner.

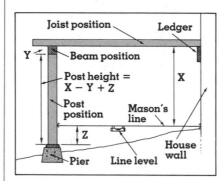

Setting posts. Set each post upright on its pier and position it on the wood nailing block or in the post anchor. Make sure the post is plumb by checking it with a level on two adjacent sides. Nail it to the nailing block or post anchor. Making sure the post is still plumb, nail the braces to it. Now you're ready to install the beams.

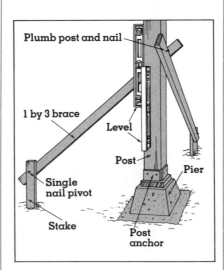

Securing beams. There are several ways to anchor beams in place (see "Four beam connectors" at right). The simplest (but weakest) method is toenailing, normally used for small ground-hugging decks. Your deck or overhead will be stronger if you use post caps, beam clips, or framing clips (many building codes require the use of metal connectors).

If the ends of the posts don't fit tightly against the beam, shim the posts with wood shingles—just wedge the shingles in place and secure the beam.

Four beam connectors

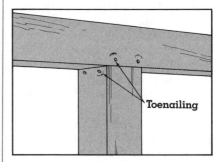
Toenailing

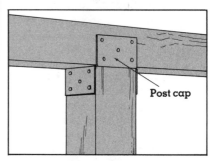
Post cap

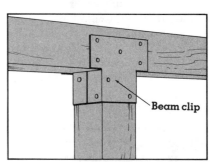
Beam clip

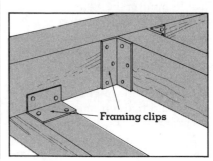
Framing clips

Bracing for posts and beams. For decks and overheads less than 12 feet high, normally only outside posts on unattached sides need bracing; check with your building department for any exceptions.

Bolt or lag screw braces to posts. Though the bracing method shown below is the preferred method, you can cut the ends of 2 by 4s or wider material at 45° angles across the width, and nail them flat to the underside of the beam and the side of the post.

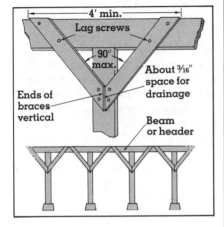

Joists and rafters

Joists and rafters serve the same function—support. Joists are used in decks, rafters in overheads. To install rafters, use the method described below for joists. (In post-and-beam construction—see following section—neither joists nor rafters are used.)

To attach joists to a ledger or a beam, you can either toenail them, support them with a 1 by 2, or set them in joist hangers (see illustration following). If the ledger or beams

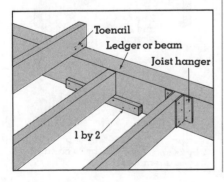

are wider than the joists, the 1 by 2 strip works well. Steel joist hangers provide the most secure attachment.

Position a joist at each side of the deck first. Make sure the joists are perpendicular to the ledger and the beam (or to the two beams, if the deck is freestanding) by using a carpenter's square or the 3-4-5 method described on page 71. Working from either outside joist, measure, set, and fasten the other joists. You may have to set the last joist closer than planned to the outside joist, so as not to exceed the maximum allowable joist spacing.

Bracing for joists or rafters. To prevent joists or rafters from twisting or bowing, brace them with staggered wood spacers, wood or metal bridging, or headers (see illustration below). Use a single row of spacers or bridging for 6 to 10-foot spans, two rows for 10 to 20-foot spans.

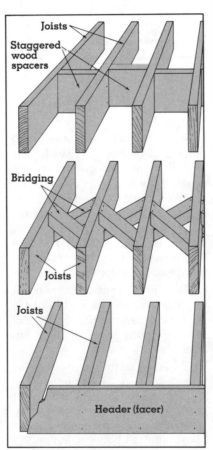

Staggered wood spacers are the easiest to install and, for appearance, the best choice for bracing

rafters. Simply cut and fit them, then secure them by nailing through rafters or joists.

Wood bridging requires more work because you have to miter the ends. Bridging is toenailed to joists. You can also use metal bridging, available in several varieties.

If joist spans are less than 6 feet, headers nailed across joist ends are adequate.

Post-and-beam construction

An alternate construction method for decks and overheads uses no joists or rafters. Instead, the beams directly support the decking or roofing. Depending on the design, this method may require less lumber than the method using joists or rafters.

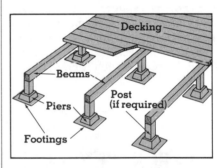

A post-and-beam deck or overhead can be freestanding or attached to a house. If attached, the beams are connected to the ledger in the same manner as joists or rafters. For a ground-hugging deck, beams can be attached directly to the piers.

The techniques discussed in this chapter for setting foundations, erecting posts, attaching ledgers, installing beams, and bracing posts also apply to post-and-beam construction.

The spacing of piers depends on the size of beams (usually piers are 4 feet apart when 4 by 4s are used). The size of the planking determines the spacing of beams—up to 9 feet when 2 by 4s are used on edge. Check with your building department for requirements that pertain to post-and-beam construction.

Decks & overheads: Decking

After you've built the deck substructure, you'll be ready to place, space, and nail the decking. One of your goals in laying decking is to make sure the decking pattern (see below) comes out even.

Arranging the decking

Place, square, and nail the first board. Then make a dry run with the rest of the boards, using a 16d common nail as a spacer—but don't nail boards to joists or beams yet. Depending on the space left at the end, increase or decrease the spacing between boards; you'll need a dif-ferent size nail, or else a wood spacer.

If the boards are too short to run the full length or width of the deck, position any joints directly over a joist or beam. Stagger joints so that no two line up consecutively over one joist. If the appearance of the lumber permits, place boards bark side up to minimize checking and cupping. The bark side is the outer or convex side of the growth rings (see illustration below).

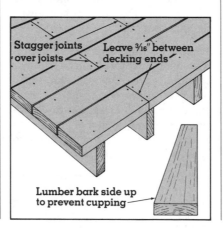

Stagger joints over joists · Leave 3/16" between decking ends

Lumber bark side up to prevent cupping

Nailing

To fasten decking, galvanized ring or spiral shank nails are preferred. You can use galvanized common nails, but they don't hold as well and you may have to reset them from time to time. With 1½-inch decking, use 16d nails. To prevent board ends from splitting when you nail them, drill pilot holes that are slightly smaller than the diameter of the nails.

Nail decking boards at every support point (joist or beam). If decking is laid flat, use two nails for 2 by 4s and 2 by 6s, three nails for 2 by 8s or wider. The job will look neater and more professional if you keep nails in a straight line. For maximum holding power, drive nails perpendicular to the surface of the board, except when you're securing an end that will have another board butted against it. In this case, drive nails at 45° angles through the board and into the supporting member.

On-edge decking should always be toenailed.

Decking patterns

Because decks are designed from the top down, one of your first decisions will involve selecting a decking pattern. The pattern you choose may affect how the deck's substructure is built. Generally, the more complex the decking pattern, the more complicated the substructure must be. To avoid a complicated supporting structure, choose one of the parallel patterns shown at right, using boards of the same or varying widths. (Two-inch lumber is recommended for decking; one-inch lumber cannot support the necessary weight.)

The simplest, soundest, and most economical decking patterns are those made of 2 by 4 or 2 by 6 lumber laid parallel and evenly spaced, running the full length or width of the deck. Whether the decking runs across the length or width depends on the deck's substructure; decking should be laid perpendicular to joists (or perpendicular to beams, in post-and-beam construction).

The simplest variations of parallel deck patterns make use of two or more different widths of 2-inch lumber. Many combinations are possible.

On-edge patterns are created when 2 by 3s or 2 by 4s are laid on their sides (on edge), usually directly on beams. On-edge decking is expensive and heavy but can span long distances between supports—an advantage if you want to eliminate the clutter of joists on a high-level deck or if you want a simple substructure under a low-level deck.

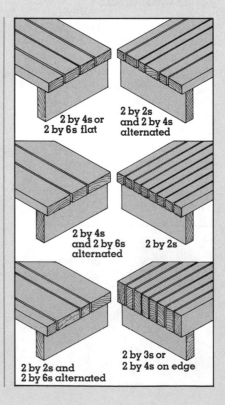

2 by 4s or 2 by 6s flat

2 by 2s and 2 by 4s alternated

2 by 4s and 2 by 6s alternated

2 by 2s

2 by 2s and 2 by 6s alternated

2 by 3s or 2 by 4s on edge

How to build a deck

Shown below in sequence are the major steps in building a simple, low-level deck attached to a house wall. This page should not be your sole reference for deck building. Rather, it is meant to give you an overall picture of the process and serve as a guide to the more detailed information in this chapter.

Before you start building, make sure the site is properly graded (see page 34) and has good drainage (see pages 48–49). Drainage problems are more likely to occur on flat lots than on sloping or hillside lots (on the latter, though, you may have to protect the slope against erosion). Clear the ground of weeds and other growth.

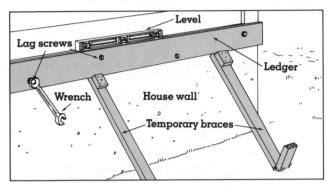

Determine position of ledger and prop it into position. Drill holes for lag screws and fasten ledger in place, making sure it's level.

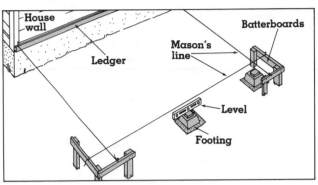

Build batterboards level with ledger top at outside corners; then locate deck edges with mason's line. Set ready-made piers on concrete footings, and level.

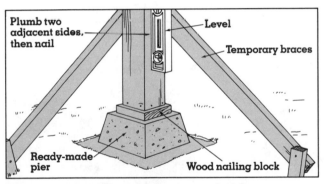

Measure posts and cut them to proper height. Temporarily brace them in place, and plumb before fastening to piers.

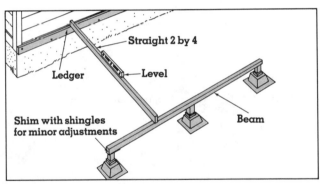

Place beam across post tops; level it with ledger, or allow outward slope for drainage. After leveling across post tops, fasten beam to posts.

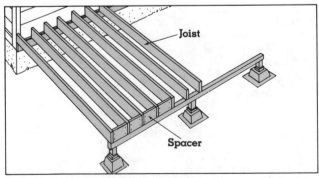

Position joists at correct intervals on ledger and beam, and fasten them in place. Brace joists with spacers at ends and, if needed, at midspan.

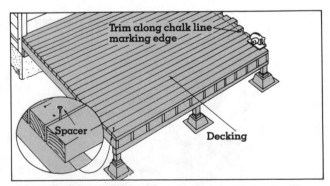

Align decking across joists, staggering joints (if any). Space boards evenly. Nail into place with 16d nails; trim edges with a power saw.

How to build an overhead

Building an overhead structure is very similar to building a deck, except that you'll probably spend a lot more time on a ladder. The illustrated steps below are intended to give you a general idea of the building sequence for a freestanding overhead. Before you begin, review the information about foundations, framing techniques, and deck construction in this chapter.

If you're not building your overhead over an existing patio, make sure the ground is properly graded (see page 34) and has good drainage (see pages 48–49). Then put in the post footings and paving (if used). If you're paving the site with concrete, it's simpler to pour the footings and the paving together. For more on paving, see pages 32–47.

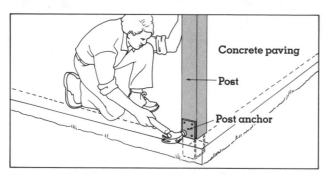

Set post in post anchor embedded in concrete footing, after cutting post to length and nailing post cap to top. Hold post vertical and nail anchor to it.

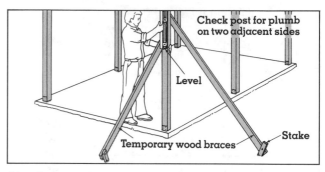

Plumb post with level on two adjacent sides; secure in position with temporary wood braces nailed to wood stakes driven into the ground.

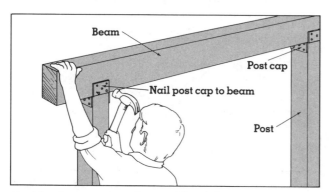

Position beam on top of posts. Check that posts are vertical and beam is level (shim, if necessary); then nail post caps to beam.

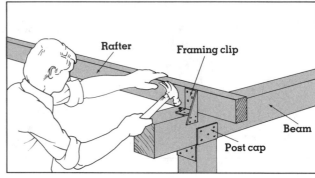

Set and space rafters on tops of beams and secure them with framing clips (shown) or by toenailing to beams. If span warrants, install bracing.

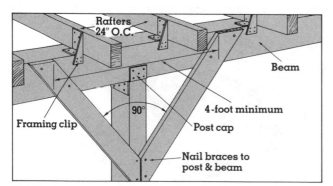

Nail or bolt 1 by 4 or 1 by 6 braces, with ends cut at 45°, between beams and posts. Cut them long enough so that ends are at least 2 feet from post caps.

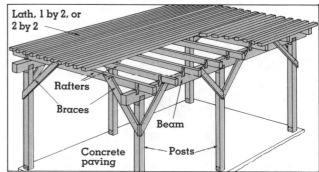

Cover rafters with lath, 1 by 2s, or 2 by 2s spaced to achieve desired amount of shading. Nail roofing material to rafters.

Roofing materials for overheads

When you select roofing, consider climate, cost, construction methods, and maintenance. And be sure the roofing you choose will create the environment that you want. One material may turn a patio into an oven because of restricted air circulation. Another, though it provides welcome shade in summer, may unduly darken an otherwise sunny room in winter.

For a solid, weatherproof roof, conventional wood and composition shingles are hard to beat. Other less conventional materials can also be used to striking advantage (see illustrations below). More information is available on these and other materials from manufacturers, your landscape architect, or your local building supply center.

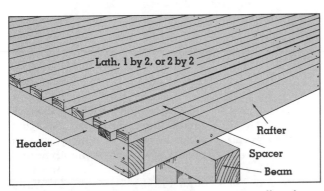

Spaced lath, 1 by 2s, or 2 by 2s are easy to install and allow you to control the shading effect by means of spacing.

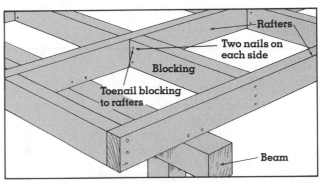

Eggcrate is open to sky but substantial enough to give a sheltered feeling. You cover it with material such as shade cloth, or grow vines on it.

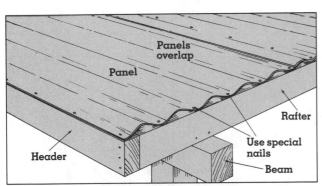

Corrugated plastic panels are often used for patio overheads. Most popular panel is 26 inches wide and, with a 2-inch overlap, fits a 24-inch rafter spacing.

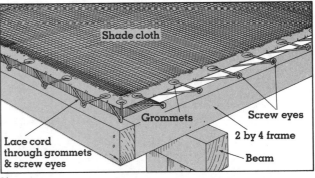

Shade cloth is available in a number of different weaves that provide 20 to 90 percent shade. Laced to a frame, shade cloth can be removed during winter months.

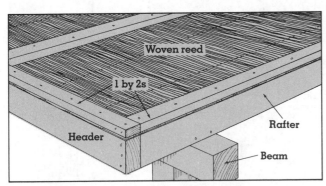

Woven reed is easy to install on roof structures. Interesting texture provides high degree of shading, and reed can be removed during winter months.

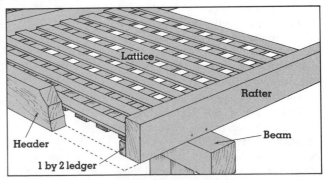

Lattice panels, set on ledgers attached to rafters, can be purchased in a number of standard sizes, or you can make your own from lath in the pattern of your choice.

PROJECT IDEAS

A varied collection for the do-it-yourselfer

This chapter is brimful of ideas for building projects that will add beauty, comfort, or convenience—if not all three—to your garden or patio. Ideas range from cold frames and hotbeds to raised beds and greenhouses; from ways of shading tender plants to ways of fending off animals that would love to munch on them; from simple garden seats to grand gazebos to garden pools.

We present these projects in idea form—without step-by-step instructions, but with clear illustrations and directions that we believe will meet the needs of the do-it-yourselfer who has basic carpentry skills. A few of the projects are more complex, requiring routing and cutting dadoes, grooves, and lap joints; some require steel work, which you can have done by a steel fabricator.

The gazebos are among the most demanding of the project ideas, and so are the pools, for which you'll need some knowledge of plumbing and masonry—you'll find masonry techniques in the "Paving" and "Walls" chapters of this book.

For any major project, check with your building department about code requirements and whether or not a permit is necessary.

Plant displays

With the plant pedestals and benches shown on these two pages, you'll be able to show off your container plants to very best advantage. Raising them closer to eye level makes plants better to look at and easier to care for. The smaller display pieces are especially attractive when several are built to varying sizes and heights, then grouped (see illustration at right).

You can alter the dimensions of any of these pedestals or benches to achieve whatever effect you want. Because they'll be in contact with the ground, make them from wood that resists rot and insects—redwood, cedar, or pressure-treated lumber—and use galvanized fastenings.

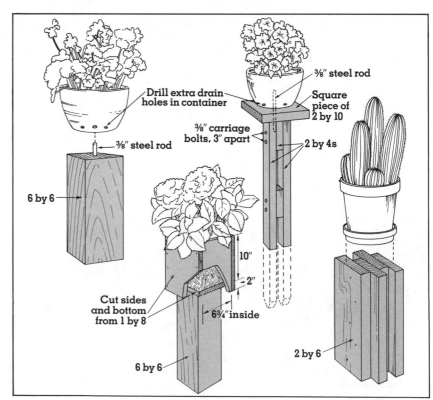

Drill extra drain holes in container

⅜" steel rod

Square piece of 2 by 10

⅜" carriage bolts, 3" apart

2 by 4s

6 by 6

⅜" steel rod

10"

2"

Cut sides and bottom from 1 by 8

6¾" inside

6 by 6

2 by 6

Pedestals for plants can be placed anywhere for seasonal displays. They're best used in groups of the same design, in varying heights.

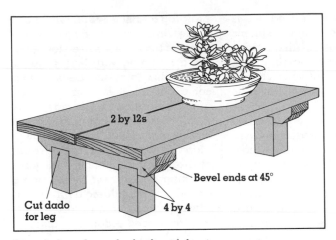

Simple bench *can be built with basic carpentry tools to any width, length, or height that suits your plants. Design: Esta James.*

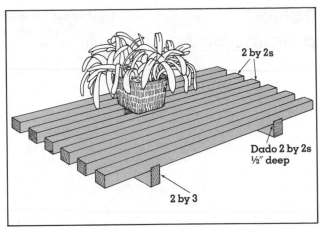

Open bench *raises potted plants off ground, keeping them free of splashed mud and putting distance between plants and soil pests.*

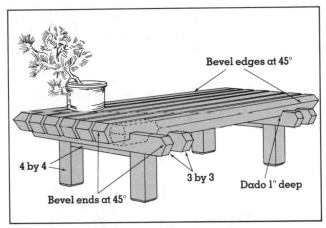

Rugged plant display table *is sturdy enough to support several heavy potted plants. Design: Lloyd Bond & Associates.*

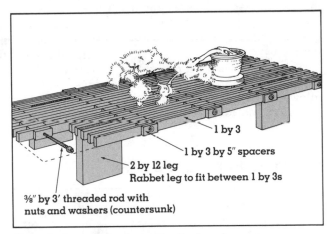

Slatted plant table *elegantly displays bonsai, rare cactus, or other specimen plants. Design: Joseph Yamada.*

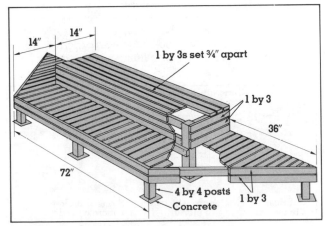

Bilevel bench *can turn narrow side yard into container plant display area. Elevating plants makes grooming easier.*

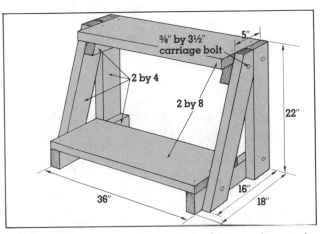

Display rack *for container plants can be moved around patio or deck to take advantage of sun or shade. Alter the size to suit your needs.*

Raised planting beds

Building a raised bed is the kind of garden project that's well worth the expense and time it takes. Well designed, a raised bed has a strong architectural value and introduces into the garden interesting color and texture in wood, stone, brick, adobe, or other materials. It can also provide a smooth transition from one level to another. And there's another advantage—your plants will be easier to care for, because you won't have to bend as far.

If you plan wisely, a raised bed will display plants impressively and allow you to grow plants that otherwise might not favor your garden.

Perfect for plants. A raised bed is perfect for dwarf or miniature plants that often are lost in the garden among taller plants. Raising them closer to the eye gives you a chance to enjoy their diminutive charm to the fullest.

Because the soil in a raised bed quickly absorbs the heat of the sun, vegetables and berries planted there will produce earlier, extending the growing season. And if you're interested in hybridizing plants, growing your specialties in a raised bed will allow you greater control of growing conditions. You can even build a temporary cold frame (see pages 84–85) on the raised bed to protect those plants in cold weather.

In gardens with heavy, poorly drained soil, a raised bed may be the only place to grow daphne, citrus, gardenias, and other plants that don't like to get their feet wet. The warm, well-drained soil of a sunny raised bed is favored by succulents, cacti, and other plants from warm, dry climates.

Building a raised bed. A raised bed is like a low-level retaining wall, and many of the methods used for building a retaining wall (see page 63) apply also to raised beds. Under some conditions, the restrictions that apply to retaining walls may apply to raised beds too. If your raised bed will be higher than 3 feet or if you're planning to build it on sloping or unstable ground, check with your building department before going ahead. When building a raised bed of wood, use insect and decay-resistant redwood, cedar, or pressure-treated lumber.

Drainage is most important. If the bed is open to the ground at the bottom, most excess water will drain out. If it's closed, make weep holes 2 to 3 inches up from the ground; space them 2 to 3 feet apart. It's also a good idea to place at least a 4-inch layer of crushed rock in the bottom of the bed before you fill it with soil. Illustrated below and on the facing page are some raised beds that require only basic building skills.

Simple-to-build raised beds

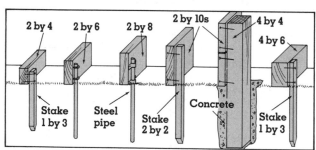

Rough-finish redwood or cedar planks make durable raised beds; stake as shown, according to plank size. Surfaced edges, square ends give best alignment.

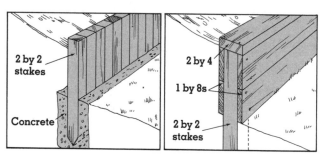

Wood stakes in a row can be driven into the ground or embedded in a narrow trench filled with concrete. Add a facing and cap of lumber for a finished appearance.

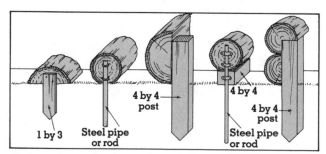

Redwood or cedar logs offer variety in shaping a raised bed. These logs are particularly suited for raised beds in natural landscaping.

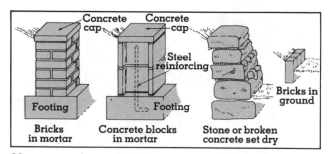

Masonry makes an enduring raised bed (see "Walls," pages 50–63, for building information). Drainpipes or weepholes will relieve water pressure.

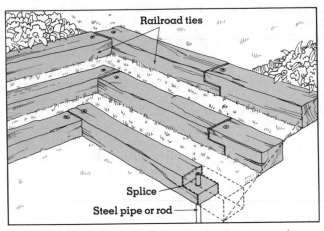

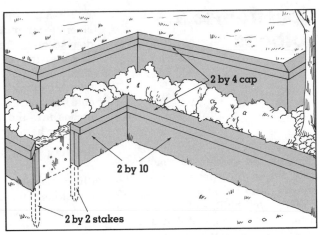

Stepped beds *made of railroad ties or 4 by 4s provide an attractive and practical transition between two levels in the garden. You can do the same thing with masonry.*

Stepped raised beds *made from 2-inch-thick redwood, cedar, or pressure-treated lumber can tame a long slope into a series of useable terraces.*

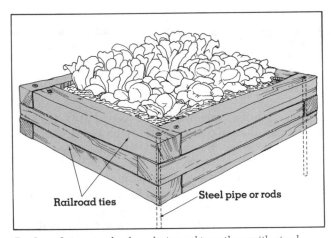

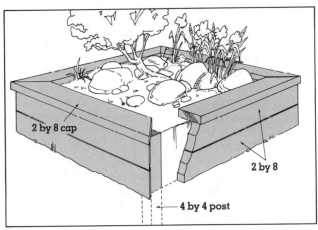

Railroad ties *stacked and pinned together with steel pipes or rods driven into the ground make ideal raised beds for a vegetable garden.*

Showcase for a small tree, *raised bed like this could also prevent roots of a larger tree from being exposed if you excavated for lower grade.*

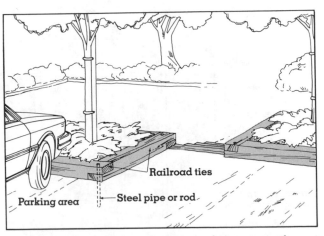

Patio paving *laid in a grid pattern with wood dividers is much more pleasing if occasional squares or rectangles contain plantings in raised beds.*

Protect shrubs *and other plants in parking area with raised beds made from railroad ties or other heavy timbers. Secure them with steel pipes or rods.*

Cold frames & hotbeds

Experienced gardeners consider a cold frame indispensable to successful year-round gardening. To many, a well-built, well-tended cold frame is nearly as useful as a small greenhouse. And it takes up far less space and can cost much less.

If the unit has an auxiliary heating system, the structure is commonly known as a "hotbed." The term "cold frame" is used exclusively here, but it applies to both cold frames and hotbeds.

Planning your cold frame. To gain maximum solar heat, orient your cold frame to face south or southwest. Build it so the cover will have a 6-inch back-to-front slope to trap the most heat inside and let rainwater run off. Whatever location you choose for your cold frame, make sure it's in a part of the garden that has good drainage, since you don't want the frame to be sitting in a puddle after every rain.

When planning its location, consider the microclimates in your garden. A stand of trees will deflect wind and keep the cold frame warmer inside. The wall of a nearby building will reflect additional heat that the cold frame can absorb. Sink it 8 to 10 inches into the ground, and you'll increase heat retention even more.

Building hints. In planning a permanent cold frame, such as the one illustrated below, start with the dimensions of the cover. You have two choices: you can make the cold frame to fit a cover you already have or can obtain inexpensively (a window sash is ideal), or you can choose inside dimensions that fit a multiple of the dimensions of planting flats. Most nurseries now use plastic trays that hold six 6-pack planters. The trays measure about 11 by 21 inches. The most common sizes of wood flats (now seldom used) are 14½ by 23½ inches and 18½ by 18½ inches.

For the cover, consider glass, acrylic plastic, fiberglass, and polyethylene sheeting—all readily available materials. You can also buy ready-made sash or snap-together aluminum sash in which you install polyethylene film. These are available in a variety of sizes.

For the frame, the material that's easiest to work with, most commonly available, and least expensive is wood. Decay-resistant redwood, cedar, and pressure-treated lumber are ideal.

Other equipment is needed, too. A good thermometer is essential if you want plants to thrive in your cold frame. Most plants that will grow well outdoors in North America will continue growing at temperatures from about 40° to 100°F/4° to 38°C, and will do best at about 85°F/30°C.

When your thermometer reaches 85°F/30°C, you can prop open the cover to let out some heat. Then in late afternoon, when the outside temperature starts to fall, shut the top to trap the heat radiated by the soil. You'll probably need to open your cold frame on all but the coldest, cloudiest days.

To help your cold frame absorb heat most efficiently, paint the sides black, inside and out. Placing a thermal mass (something that holds heat well, such as glass or plastic gallon jugs filled with water) inside the cold frame keeps night temperatures higher. If the jugs are painted black, the water inside them can reach 135°F/57°C, an effective temperature for heating.

On very cold nights, cover the cold frame with carpet, straw, or sheets of styrene foam to keep the heat in.

A close look at a cold frame

Cold frame cover acts as a heat trap; wall behind reflects sunlight, shields against winter winds. Notch transverse mullions for runoff. To control temperature and humidity, use appropriate lengths of wood as props.

Nine portable cold frames

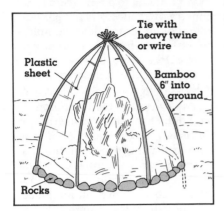

Bamboo or other flexible stakes make frame for small tent of plastic sheet. Anchor plastic with rocks.

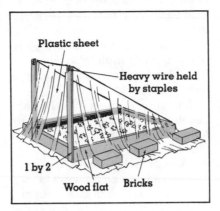

Two supports nailed to one end of wood flat hold piece of clear plastic sheet. Bricks hold plastic in place.

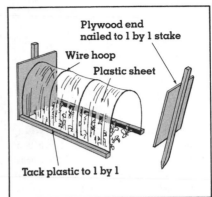

Stakes hold plywood ends in place. Ends of stiff wire hoops pushed into ground support plastic sheet.

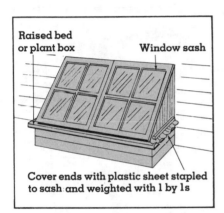

Hinged sash attached to side of house covers raised bed or plant box. Prop sash open for ventilation.

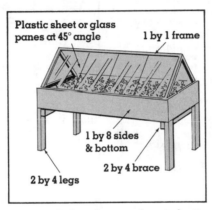

Elevated plant box can have plastic or glass cover. Board up ends or cover them with plastic sheet.

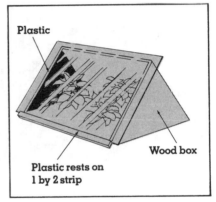

Wooden box cut diagonally from end to end has one side removed. Box supports plastic at a slant.

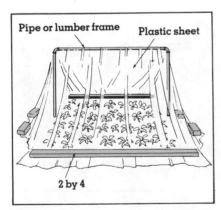

U-frame of pipe or lumber stuck into ground supports tent of plastic sheet. Weight edges with wood strip.

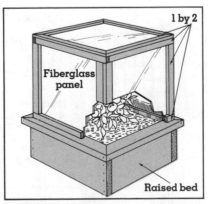

Miniature hothouse prolongs growing season. Fiberglass panels are framed in wood.

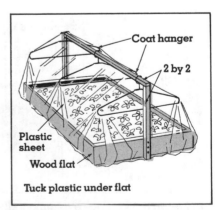

Coat hangers hung on 2 by 2 supported by vertical 2 by 2s form tent frame for plastic sheet.

Trellises & frames

There are many reasons why a young shrub or vine needs more support than a stake can give. The cross-barred structure of a trellis offers more width and more places to tie a sprawling plant. When a plant is splayed out in graceful curves and masses against a background grid, it often looks better and displays its flowers or fruits better than when it's allowed to grow freely in all directions.

A framework that holds a plant away from a wall also keeps it healthier. Good air circulation minimizes mildew and rot.

Rampant growers like ivy can benefit from the discipline of a sturdy trellis. A strong framework offers the vine some alternative to filling every crack and crevice in a fence or wall, and it encourages a gardener to keep the vine trimmed back. (Almost any plant needs more pruning and pinching when grown on a trellis than when grown elsewhere.)

Finally, the trellis itself adds year-round interest to a fence or a blank house wall, especially if it provides some color contrast—for example, a dark-stained framework against a light-colored wall (or vice versa), or glistening white latticework set into a dark green fence.

The examples below and on the next page show the great variety of trellises you can build. Build them with care, using good materials and strong galvanized nails, bolts, and screws. Growing plants can exert a lot of force (ivy can destroy a fence), and the weight of a mature plant can be much greater than one might expect.

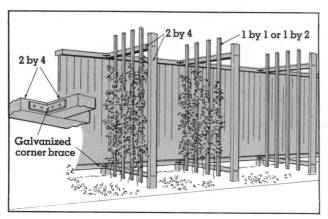

Wing trellis gives best effect when several are attached to fence. Attach 2 by 4 frame to fence with angle bracket, setting vertical 2 by 4 in concrete. Nail 1 by 1s or 1 by 2s to horizontal 2 by 4s.

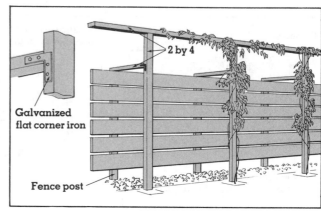

Open frame of vertical 2 by 4s set in concrete and capped with 2 by 4s is set out from existing fence and attached to it with 2 by 4s. Height is limited only by local building code or zoning laws.

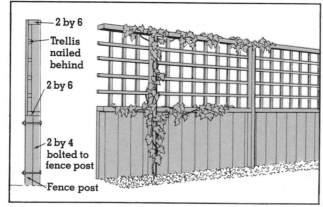

Add height to low fence by setting trellis into frame attached to fence. To make frame, bolt 2 by 4 uprights to fence, nail 2 by 6s to top of fence and top of 2 by 4s, and nail trellis to them.

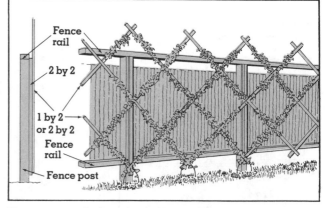

For easy espalier, nail upright 2 by 2s to faces of fence posts. Then nail 1 by 2s or 2 by 2s in diagonal pattern to upright 2 by 2s. Paint diagonals a contrasting color and train plants along them.

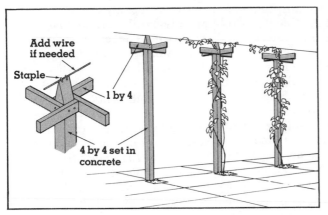

Stylized tree *made from 4 by 4 set in concrete and with 1 by 4 branches can display grape, wisteria, gourd, climbing rose, or honeysuckle vines. Make 4 by 4 any height, but don't make 1 by 4s longer than 18 inches.*

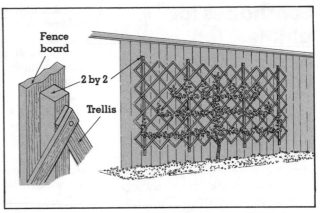

Espalier frame *is prefab trellis mounted away from fence to allow circulation around plant. Nail upright 2 by 2s to fence boards, clinching over nails in back. Then nail trellis to upright 2 by 2s.*

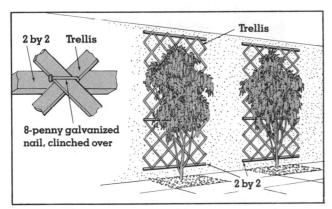

Prefab sections *of trellis, available in several sizes, are attached to 2 by 2s securely fastened to fence or wall. Attach trellis by driving galvanized nails into 2 by 2s and clinching ends over trellis.*

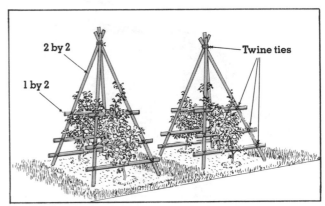

Pyramid for tomatoes *provides good support and air circulation and shades root area. Set 2 by 2s about 6 inches into ground and tie tightly at top. Tie horizontal 1 by 2s to the 2 by 2s as illustrated.*

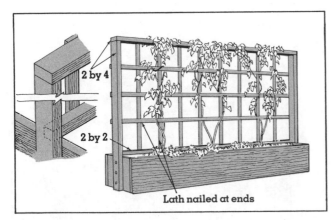

Freestanding screen, *made from a 2 by 4 frame and lath, is attached to a simple planter and covered with vines. It makes a good privacy screen, space divider, or windbreak. Mount it on heavy-duty casters for mobility.*

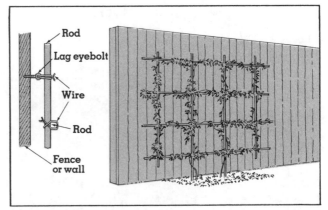

Strong trellis *of steel reinforcing rods is built by setting lag eyebolts into wall or fence and forming grid of rods wired to each other and to eyebolts. In hot climates, steel becomes hot and may harm plants.*

Sunshades for plants

Seedlings and young transplants—even those of sun-loving plants—get a much better start if some temporary shade is provided. This is particularly true in hot-summer climates, but in almost any region a sudden surge of hot weather can damage young plants. Some full-grown plants are shade-loving by nature—for example, tuberous begonias, cinerarias, cyclamen, and fuchsias. In all these situations, sunshades are the answer.

Look for ready-made material when building sunshades. Lath-and-wire fencing, bamboo or plastic blinds, inexpensive reed fencing, and shade cloth all work well.

Shade cloth used by professionals (usually black or green and made out of polypropylene or saran) is widely available. Choose shade cloth with a density that permits a soft, even light to reach your plants. The size will depend on the dimensions of the support structure.

A lath sunscreen should be placed so that the laths run north and south. (As the sun moves across the sky, strips of sun and shade move across the plants.) Also, since the sun is always a little to the south, let the south edge of the sunscreen extend beyond the plants to keep them all in shade.

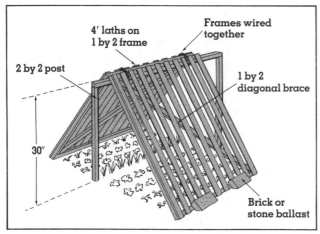

Tent shelter *is supported by 2 by 2-inch wood members. Two lath-covered frames are wired together at the top, and have diagonal braces of 1 by 2.*

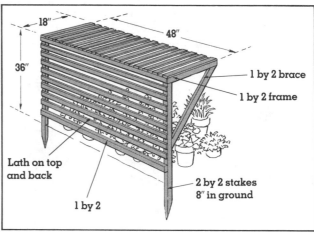

Lightweight sunshade *can be moved to accommodate changes in sun's direction as seasons change. Brace adds to roof's stability.*

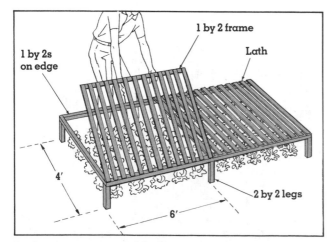

Lath screen *on a portable frame can be placed over beds of new seedlings as shown, or can be propped up on south side of planting bed.*

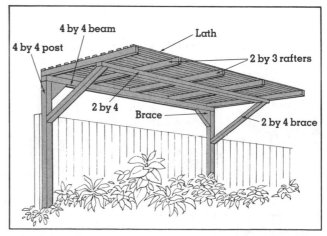

Attached to fence, *lath overhang gives protection to shade-loving trees and plants. In place of lath, you could use shade cloth or a bamboo shade.*

Animal barriers

Allow rabbits, deer, birds, and other wild creatures free rein in your garden and they'll soon gobble up everything they consider edible—and that includes your fruits, vegetables, and flowers.

It takes some ingenuity to design protective barriers for your garden, keeping in mind both the crop and the creature you're trying to protect against. A high and sturdy fence may keep out deer, but rabbits and birds will still have free access. Most of the ideas shown below were designed for a specific purpose. You can adapt them, if needed, to design a barrier to protect your garden from these hungry critters.

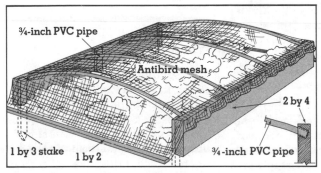

Antibird mesh spread over plastic pipe arches is held in place by 1 by 2s attached to ends. Mesh can be lifted to one side for access to strawberries.

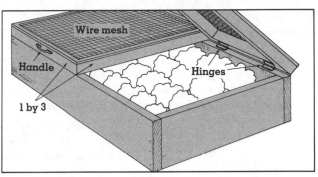

Screened lids with handles protect crops in raised bed from birds and deer. Hinged lids are easily raised for cultivating and picking.

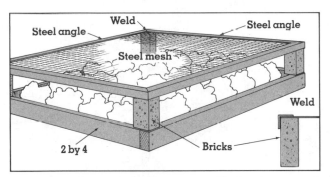

Custom-made metal grates supported by bricks discourage deer. To accommodate growing plants, stack bricks at corners, raising grates.

Lower part of wire-mesh fence is buried in ground to discourage burrowing animals such as rabbits and gophers. Embed wire in concrete for added protection.

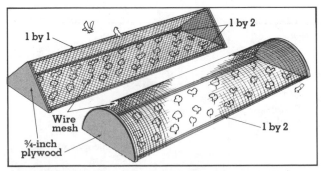

Portable protection against birds can be made to any size you need. Use scrap lumber plus chicken wire, cheesecloth, or antibird netting.

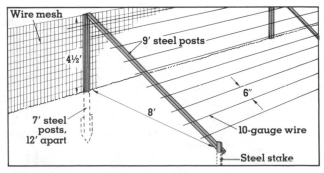

Deerproof a fence by covering it with wire mesh and building a wire outrigger extending 8 feet from outside of fence.

Water for your garden

A permanent irrigation system in your garden will save you untold hours of work and help keep your plants in good health. You have a choice of two kinds of systems: sprinkler and drip. A sprinkler system uses high water pressure and volume to disperse water over a large area, such as a lawn. A drip system dispenses water at low pressure and volume to specific areas—often to individual plants.

Planning. The first step is to make a scale drawing (¼ inch to 1 foot) of your garden, showing the locations of buildings, walks and driveways, patios and decks, water meter, hose bibbs, the water supply pipe to the house, and all the plantings you'll want to water.

The actual design of the system depends not only on your requirements but also on the types of heads you select. Manufacturers' literature usually contains all the information you'll need to make the selection and to design the actual system. You'll need to mark on your drawing the locations of sprinkler heads or drip emitters, pipes and pipe fittings, an antisiphon control valve for each branch, and the main shutoff valve for the system. Check with your building department about whether or not you need a permit.

Water pipes and pressure. To ensure adequate water at adequate pressure, connect the irrigation system directly to the water supply pipe (1-inch diameter or larger is best) that serves the house, on the house side of the water meter. Regardless of the size of the supply pipe, don't use a pipe smaller than 1 inch to supply your system.

Because drip systems require only low water volume and pressure, you may be able to connect a small drip system to a convenient outdoor faucet; the manufacturer's literature will help you determine whether or not this is best for your situation. If you have a water softener, by-pass the water to the outdoor faucet.

How to install a sprinkler system

To assemble a sprinkler system (see illustrations below) you'll need the following: a supply of PVC pipe and fittings, PVC pipe-cleaning compound and solvent, a main shutoff valve, an antisiphon control valve for each branch, galvanized or plastic pipe for sprinkler risers, and sprinkler heads.

Several kinds of sprinkler heads are available, including spray heads

Installing a sprinkler system

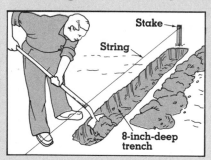

Dig 8-inch-deep trenches for pipes. To keep trench lines straight, tie string between two stakes.

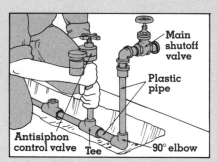

Connect antisiphon control valve to water supply; set it at least 6 inches above ground.

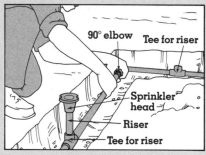

Assemble pipe, from control valve outward, fitting risers and heads to tees as you move along.

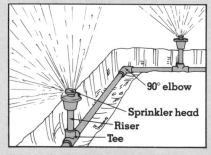

Test for leaks and coverage after welds are dry, having first flushed out pipes with heads removed.

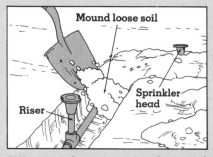

Fill in trenches, mounding loose soil above ground level along center of each trench.

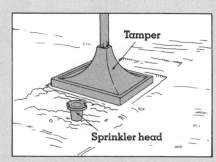

Tamp soil firmly along length of trenches, using a heavy tamper to minimize settling.

with quarter, half, three-quarter, and full spray patterns; impulse or cam-driven heads; and bubbler heads.

PVC pipe comes in 10 and 20-foot lengths. If you need shorter ones, measure carefully and cut with a hacksaw, making the ends square; remove any burrs with a file.

When you're joining plastic parts, work quickly. PVC solvent cement dries rapidly, and once it dries, joints can't be broken apart. First apply the cleaning compound; then before it dries, daub the solvent on both parts, shove the parts tightly together, and give them a half-twist to make a seal.

After installing the control valves, let the welds dry for 6 hours. Then turn on the water and check for leaks.

How to install a drip system

To assemble a drip system, you'll need the following: a supply of ⅜ or ½-inch black polyethylene hose and fittings (you can buy these in a kit with instructions), a shutoff valve, an antisiphon control valve for each branch, a pressure regulator, a filter, end caps, transfer barbs, polyethylene microtubing, and emitters.

Emitters are the devices that dispense water in a drip system. They slow the flow of water from the line and dispense it drop by drop. Some types of emitters can be plugged directly into the hose and, if desired, extended with a length of microtubing; others can be plugged into the end of a length of microtubing that is connected to the hose with a transfer barb. Some emitters have adjustable flow rates, some are self-flushing, and some can be shut off except when they're needed. An emitter with a rated flow of 1 gallon per hour is usually adequate.

Other types of dispensing devices are available for use with drip systems. In-line emitters and perforated or drip tubing in a variety of types can be used to irrigate row crops. Misters, microsprayers, and minisprinklers can be connected to a drip system to create a small-scale version of a sprinkler system.

The polyethylene hose you'll use is much easier to unroll if you leave it in the hot sun for an hour to soften. When it's pliable, connect the hose and lay out the main lines, connecting laterals with tee or 90° elbow fittings. You can bury the lines in shallow (2 to 4-inch) trenches or lay them on the ground, leaving them exposed or covering them with mulch. Locate filters and hose ends so they're easy to flush.

After all lines have been laid and flushed, drill holes in the hose as the manufacturer directs, insert emitters (or transfer barbs), and cut and fit microtubing.

The system will require regular maintenance, because sand and dirt can plug microtubing and emitters. To keep the system in shape, flush the lines every 4 to 6 months, wash filter elements every month, and clean any plugged emitters.

Installing a drip system

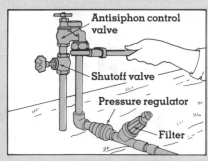

Assemble antisiphon control valve (at least 6 inches above ground), filter, and pressure regulator.

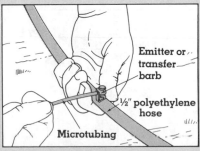

Connect polyethylene hose and lay out main lines in shallow trenches or on surface.

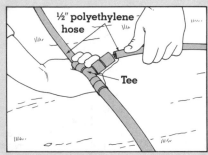

Lay out and attach lateral lines. This kind of tee has barbs that hold tubing without cement.

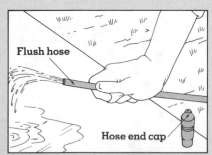

Flush any dirt out of system by running water through lines after all are assembled. Attach end caps.

Drill holes in hose; insert emitters (or transfer barbs). Cut and fit microtubing from emitters (or barbs).

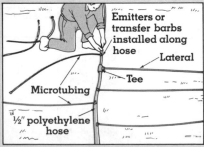

Flush again and make sure all emitters work. Cover all lines; leave ends of microtubing above ground.

Garden seating

You can buy benches and seats, but designing and building your own allows you to have garden seating that blends with your landscaping and is exactly right for you. You may want a cozy bench in some secluded spot in your garden, a wide plank atop the wall of a raised planter, or a bench that also functions as a safety barrier at the edge of a deck or steep slope. The benches illustrated on pages 94 and 95 and the curved bench on the facing page show some of the possibilities for garden seating. Below and on the facing page, you'll find construction tips and information on design and materials.

Bench design

Most people who build their own garden seating want to experiment with the design until it suits their own special needs. You don't have to stick to one basic form. For comfort and usability, though, keep the following considerations in mind:

For maximum comfort, a seat should be 15 to 18 inches high, the approximate height of most chairs. If you plan to use a thick mat or cushion, build the seat lower. Make it lower still—6 to 8 inches lower—if you intend to use it primarily for sunbathing.

There is no set guide for depth. A bench only 12 inches deep, though common, looks more like a place to perch than a place to relax. A depth of 15 to 18 inches is comfortable, and you can make it even deeper for lounging—24 inches is the width of a standard lounge pad. Or you may want to be able to stretch out on a mini-deck, like the bench for lounging shown on page 95.

Bench legs should be sturdy enough for solid support and still be in scale with the rest of the bench. Space the legs about 3 to 5 feet apart if the legs are 4 by 4s or material of similar strength. If you're using light legs such as 2 by 4s, or if the lumber for the top of the seat needs additional support to prevent it from sagging, place the legs closer together. To keep legs from wobbling, you'll need to set them in concrete, at least 18 inches deep (see page 66).

Instead of wood, you can make legs from bricks, concrete block, galvanized steel pipe (1¼ inches or larger), or clay flue pipe, or you can use steel angles, bars, and channels. All of these materials are decay and insect resistant, but most

Typical bench construction

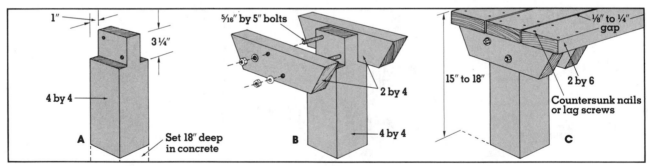

To build a simple bench, *saw shoulders in tops of 4 by 4 posts and set posts in ground (A); cut 2 by 4 braces and bolt to posts (B); nail planks to tops of braces (C).*

Building a mitered bench corner

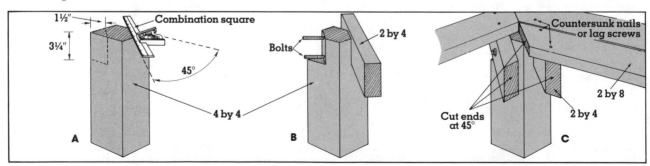

To build a bench corner, *mark 4 by 4 corner post for shoulder cuts (A); saw shoulders in top of post and bolt 2 by 4 braces to post (B); miter ends of 2 by 8 planks and nail planks to braces (C).*

take more skill to install.

Be sure to plan for a kick space underneath the seat. Not being able to tuck your feet under can be quite uncomfortable.

Choosing materials

Redwood, pine, cedar, fir, and cypress are the woods most frequently used for outdoor seating. If the wood for your bench is to be sunk into the ground or will touch the ground, choose all-heart redwood, cypress, or cedar, or use pressure-treated lumber. Builders and owners stress the importance of using the best grade of lumber for the upper surface of the seat. Well-seasoned lumber, without loose knots, will keep its shape and last much longer than a poorer grade.

For strength, use 2-inch-thick lumber for bench tops. You can also use 1 by 2 or 1 by 3 lumber if you set it on edge, as shown on this page.

Though a single wide plank such as a 2 by 12 looks stronger than two 2 by 6s or three 2 by 4s, the wide plank has a greater tendency to warp or split. If your plans call for planks 8 inches or wider, be sure to choose well-seasoned ones.

To prevent unsightly staining of wood, use only galvanized nails, screws, and bolts. For extra-solid construction that will take abuse and resist the effects of frost heave in cold climates, use bolts and lag screws to fasten the pieces together, and check them yearly for tightness.

Tips for building & finishing

Prebore the holes for the screws and bolts, and countersink them so the heads don't extend above the surface where they can snag clothing. If you're using nails, predrilling the holes will lessen the chance of the wood's splitting. Set the nail heads with a nail set.

Once the seat is built, you have the choice of leaving the wood natural and allowing it to weather, or applying a finish, such as a sealer, stain, or paint. In either case, sand or plane all exposed wood surfaces smooth and round off all edges to eliminate splinters and rough areas that might snag clothing or skin. You may need to do this periodically to keep the surfaces smooth and free from splinters.

See your paint dealer for information on the various wood finishes available. Be sure to say what the finish is for—you don't want a finish that doesn't dry completely.

Before you apply the finish, try it out on some leftover pieces of lumber. Make sure you use a piece that's large enough to give you a good idea of what the color and finish will be like.

If you want to fill the nail holes, use a nonoily filler unless you plan to paint the seat. Oily fillers discolor the wood.

Building a curved bench

You may view the idea of building a curved bench with a certain amount of trepidation. But if you work carefully according to these illustrations, you should be able to handle the job.

You'll need some bar clamps to hold the slats in place while you nail them. To avoid splitting the wood, predrill the nail holes with a drill bit that's about 75 percent of the nail diameter.

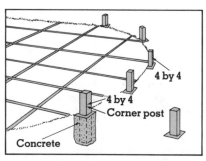

Arrange 4 by 4 posts in arc with 30-foot radius; set posts in 18 inches of concrete. Post height is 15½ inches.

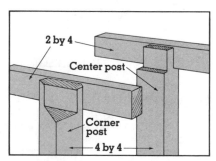

Cut post tops for two 2 by 4 braces, each 18 inches long. Bolted to post, they should protrude ¼ inch above.

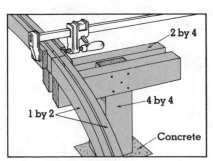

Nail 1 by 2s to braces, using clamps to hold slats in place. The ¼-inch spacers keep spacing uniform.

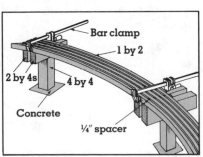

Curve slat to follow arc, nailing the slat securely to each brace. Repeat for each slat.

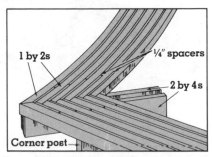

Lap slats at corner as shown, and set mitered 1 by 2 at each end to complete bench construction.

Six garden seats

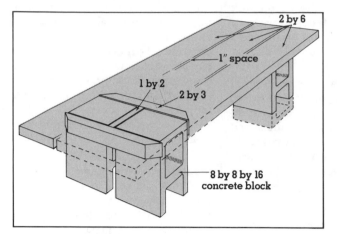

Simple bench *has H-shapes on bottom to hold it in position on concrete blocks, which can be set about 2 inches deep in concrete paving or in the ground.*

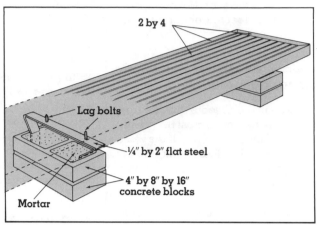

Low slatted bench *rests on U-shaped straps set in wet concrete poured into cores of concrete blocks. Bench top is made of 2 by 4s and spacers nailed together.*

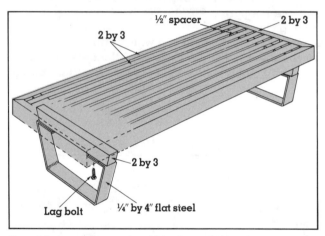

Portable bench *is made of 2 by 3s and ½-inch spacers nailed together. Formed legs of flat steel, lag bolted to cross braces, provide support for the bench.*

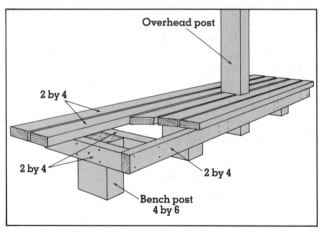

Fifteen-foot-long bench *uses post of patio overhead for support. Cross braces (2 by 4) are hidden from view by 2 by 4 fascia nailed to their ends.*

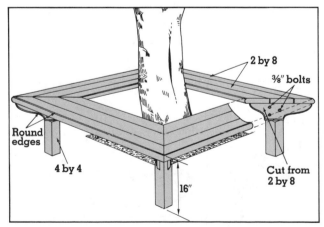

Square bench *provides a shady place for relaxation and protects tree from patio traffic. Build far enough away from trunk to accommodate tree's growth.*

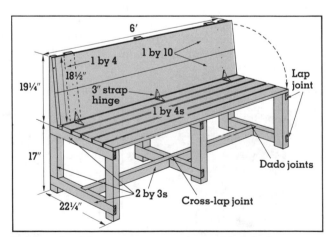

Folding back *lowers onto seat of portable bench to keep it dry and clean. Dado, lap, and cross-lap joints are best made on a table saw, using a dado blade.*

Bench for lounging

This generous bench for lounging, sunbathing, or *tête-à-tête* picnicking appears to float above the ground.

Set the 4 by 4 posts in concrete about 12 inches into the ground (below the frost line in freezing locales; information on setting posts appears on page 66). Toenail the 4 by 4 beams to the posts; shim, if necessary, to make them level. Nail the 2 by 6s to the beams and countersink the nails. Miter the ends of the fascia boards and nail them to the edges of the deck. On page 75 you'll find more information on post-and-beam construction.

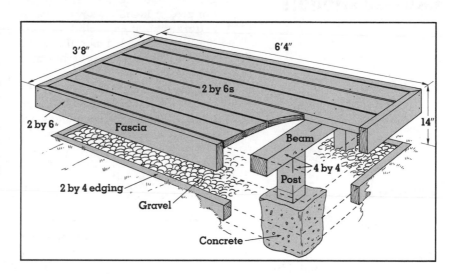

Planter barrel bench

In this project, half-barrels serve both as planters and as bench supports. You can usually find barrel halves at nurseries, lumber yards, and building supply centers. Make sure the ones you buy are sturdy. Buy one more half-barrel than the number of bench sections, so that a half-barrel will support each end.

You'll need to cut compound angles at the ends of the 2 by 6s to match the barrels. If you've never done this before, practice on scrap lumber until you have it down pat. You'll also need to cut the ends of the 2 by 3s on a curve to fit the curve of the barrels.

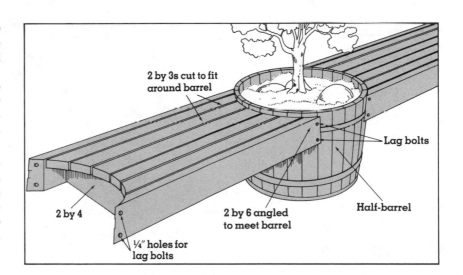

Planter seat

You can build this planter seat and simply set it on a deck or paved patio, or you can set 4 by 4 posts and 2 by 8 bench supports in concrete about 12 inches into the ground (below the frost line in cold climates). Build the two planters and position them. Build the bench frame of 2 by 4s and 2 by 8s and lag-screw to the planters. Nail the 2 by 6s in place.

For a more handsome appearance, round the edges of the 2 by 6s on the bench and planter, as well as the 2 by 4 fascia on the bench. Use a router with a ¼ or ⅜-inch-radius rounding bit.

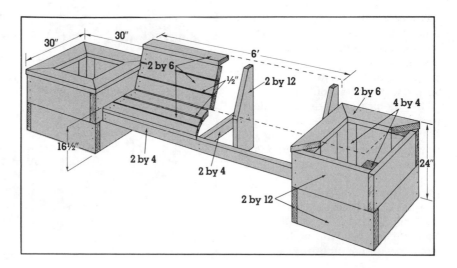

Garden shelters

On this and the next five pages are project ideas for several types of garden shelters, including gazebos. You'll need a permit, so be sure to check with your building department.

Elsewhere in this book you'll find more information that will help with these projects—see "Installing fence posts," pages 66–67, and "Decks & Overheads," pages 70–79. You can build a shelter on a deck or on an area paved with any of the materials discussed in the chapter beginning on page 32.

For the shelter, select decay and insect-resistant lumber such as redwood, cedar, or pressure-treated lumber, particularly when the wood will be in contact with or close to the ground. Be sure to use galvanized nails, bolts, and other hardware.

A-frame shelter

A quiet place for some time alone, this A-frame structure is a shelter against sun and prevailing winds. Build the framework with 2 by 8 rafters set 2 feet on centers and set 2 feet deep in concrete (below the frost line in cold climates). Notch the tops of the rafters to receive the 4 by 8 ridge beam, and nail the rafters to it. Cut gussets from 1 by 6 lumber and nail to both sides of each rafter, tight against the bottom of the ridge beam. Cover the sides with 2 by 2s spaced 2 inches apart and nailed to the rafters.

You can paint the wood, stain it, or let it weather naturally. Any type of garden paving (see pages 32–47) can be used for the floor. Design: Armand Ramirez.

Garden pavilion

Here's a garden pavilion with room for many kinds of activity. You might lounge in a swing suspended from the beams, eat a leisurely brunch at a patio table, or take a siesta on a comfortable chaise. You can enjoy the shade here on hot sunny days, or listen to the patter of a warm spring rain on the roof.

Set the posts at least 3 feet deep (below the frost line in cold climates) and in concrete (see page 66). Bolt the lintels to the posts and then nail the beams and plates in place. Cut the rafters and nail them to the plates and ridge beam, nail the sheathing in place, shingle the roof, and install the fascia and the trim. You can paint, stain, or leave the wood to weather naturally.

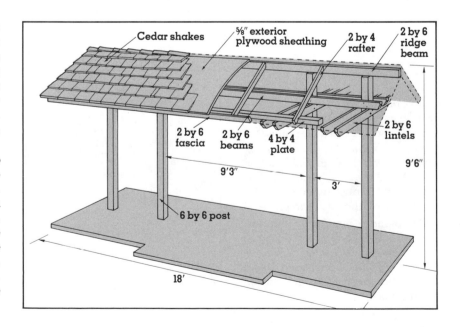

Garden pergola

Cover this simple structure with deciduous vines and you'll have a cool haven in summer and a place to enjoy the sun's warmth in spring and autumn. Until the vines mature, you can cover the pergola with shade cloth.

Assemble each post by nailing two 2 by 4s and three 2 by 4 spacers together with 16-penny galvanized nails. The spacer at the top should project 7¼ inches beyond the ends of the 2 by 4s. Set the posts 3 feet deep in the ground and in concrete (below the frost line in freezing climates); make sure the shoulders at the tops of the posts are level.

Bolt or nail the 2 by 8s to the tops of the posts. Next, set the 4 by 8 rafters in place and toenail them to the beams (if you want to miter the ends of the joists for appearance, do so before you set the joists in place). Finally, space and nail the 2 by 2s to the tops of the rafters.

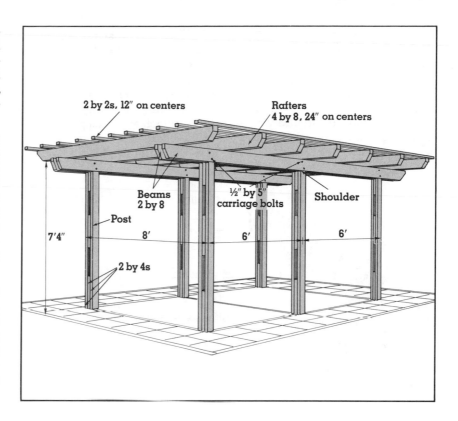

Overhead screen

At first glance, this overhead screen has a clean, finished appearance and seems lower than most overheads. A second look tells you why: 2 by 3s nailed beneath the framework instead of on top hide the usual clutter of beams and rafters.

Set the 4 by 4 posts about 3 feet deep in the ground and in concrete (below the frost line in cold climates). After setting the posts, complete any paving. Now nail the 1 by 8s and 1 by 3s to the sides of the posts. Bolt or nail the 2 by 8 beams to the posts, making sure they're level. Attach the 2 by 6 joists to the beams with hangers.

Nailing the 2 by 3s to the undersides of the joists is an awkward job at best, but here are a couple of tips: You can rent a pneumatic nail gun and air compressor and let the gun do most of the work for you. To make the job easier, drill pilot holes in the 2 by 3s before nailing them to the joists. Design: Klaus Hertzer.

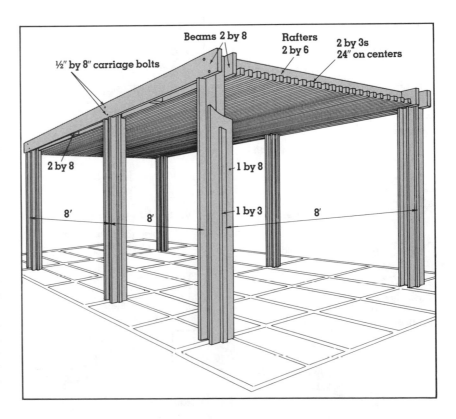

Traditional gazebo

This lattice gazebo, reflecting the charm of an earlier time, is ideal for entertaining or as a shady refuge.

You can build an octagonal deck for the gazebo, as shown, or erect it on an existing patio or deck. Build the wall panels on a flat surface; then stand them up and bolt them together, leaving the bolts at the top loose for easier assembly of the rafters later. Nail or bolt the wall panels to the deck or patio surface, too.

Bolt the long and short rafter sections together on the ground. Then, after drilling pilot and countersink holes, attach each assembled rafter to the octagonal center plug with screws. With the help of several friends, lift the assembled rafters into place, with the rafter ends between the 2 by 4 uprights. Make sure the rafters are tight against the spacers. Drill holes for the bolts and bolt the rafters to the uprights. Tighten the bolts through the tops of the uprights. Design: Gazebo Nostalgia.

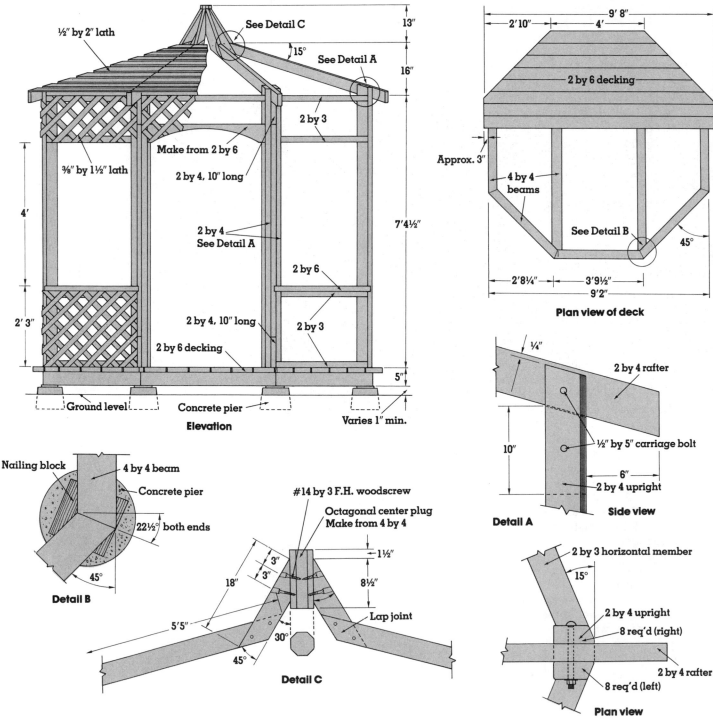

Screened gazebo

On this gazebo, screens of closely spaced boards give protection from sun and wind.

You can set the posts in concrete directly in the ground (about 3 feet—but below the frost line in cold climates); attach them to post anchors set in concrete footings; or fasten them to the substructure of a deck.

Attach 2 by 4 and 2 by 6 cross pieces, except where you want access.

Have a steel fabricator make the steel center bracket for you. Give the fabricator a short piece of the 3 by 6 rafter material so that the rafters will fit the bracket properly. Position the rafters in the bracket and use the holes in the bracket as guides to drill

½-inch bolt holes in the rafters. Secure the rafters to the bracket with galvanized ½-inch by 3½-inch machine bolts, washers, lockwashers, and nuts. Now set the assembly in place on the posts (you'll need several helpers). Bolt the rafters to the posts. Finally, toenail the 2 by 4 louvers in place. Design: Roger Fiske.

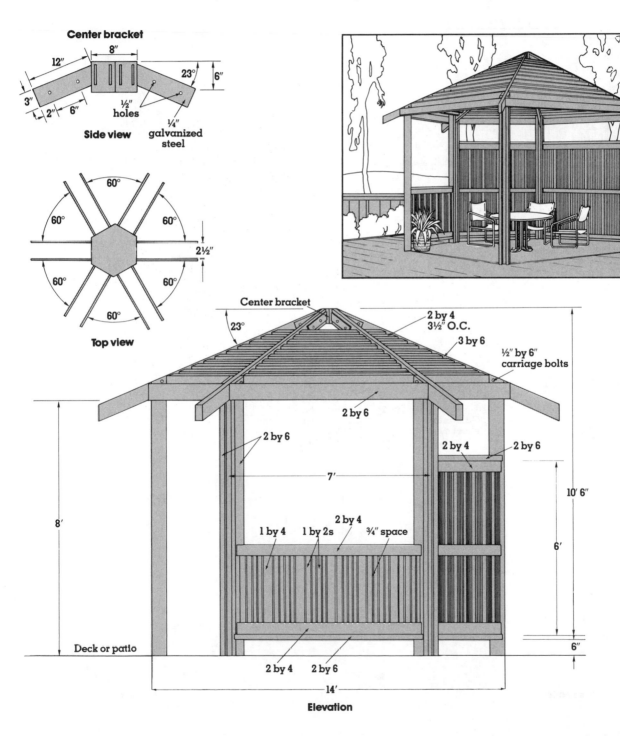

Entertainment pavilion

A shelter for entertaining or relaxing, this pavilion is as delightful as it is functional.

You can build an octagonal deck (see page 98) for the pavilion as illustrated, or you can set it on an existing deck or paved surface. The roof of 2 by 4s and ⅝-inch exterior plywood can be covered with galvanized sheet metal (as shown) or with shingles. If you choose sheet metal, consider painting the wedge-shaped sections in cheerful colors.

Canvas curtains can be pulled to shut out wind, rain, or sun. The curtains are attached to large wooden rings that slide along closet poles mounted on the inside of the pavilion. If you wish, you can make removable screens to keep flying insects out during warm weather. You may want to wire the pavilion for lights, music, and small cooking appliances. Design: Roy Rydell.

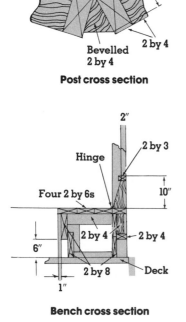

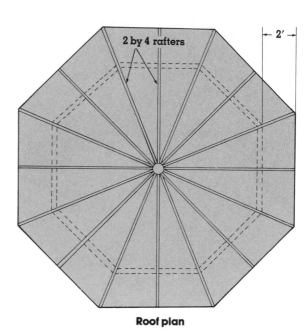

Roof plan

Post cross section

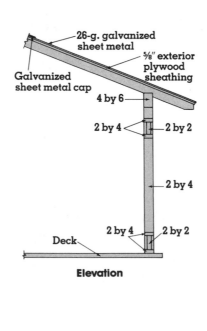

Elevation

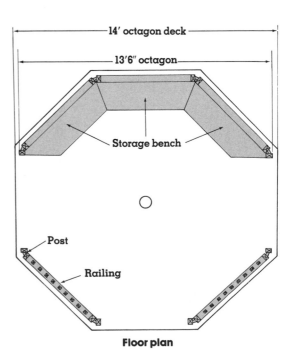

Floor plan

Bench cross section

Octagonal gazebo with storage

This octagonal gazebo, only 11 feet in diameter, is spacious enough to house two dressing rooms with built-in seats, as well as a pump house for the pool. If you prefer, you could use that space to store garden equipment or tools instead.

The open-beam roof is supported by eight 4½-inch-diameter steel columns sunk into the ground and embedded in the concrete poured simultaneously with the exposed aggregate floor. Walls and doors are ¾-inch exterior plywood. The copper ornament that tops the vent stack for the pool heater is actually a toilet tank float. Design: Morgan Stedman.

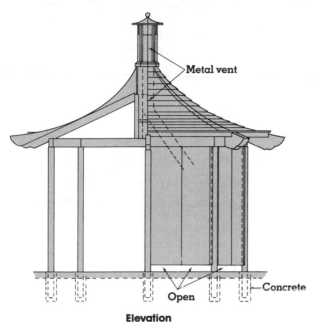

Elevation

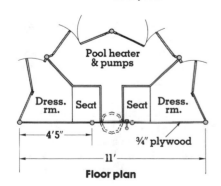

Floor plan

Roof plan area labels:
Detail B
2 by 4
8"
2 by 6
4 by 6
³⁄₁₆" steel plates welded to tops of columns
Open
Two 2 by 6s

Roof plan

Floor plan labels:
Pool heater & pumps
Dress. rm.
Seat
Seat
Dress. rm.
4'5"
¾" plywood
11'

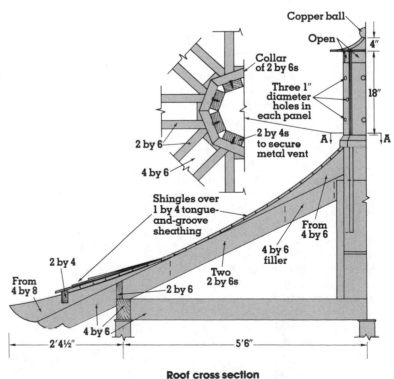

Roof cross section labels:
Copper ball
Open
4"
Collar of 2 by 6s
Three 1" diameter holes in each panel
2 by 4s to secure metal vent
18"
A — A
2 by 6
4 by 6
Shingles over 1 by 4 tongue-and-groove sheathing
2 by 4
From 4 by 6
4 by 6 filler
From 4 by 8
Two 2 by 6s
2 by 6
4 by 6
2'4½"
5'6"

Roof cross section

Lighting for your garden

For highlighting plants, discouraging prowlers, or illuminating paths, outdoor lighting is a welcome addition to any house. You can extend your home's 120-volt system into the garden to power a variety of permanently placed fixtures (be sure to check with your building department), or you can step the system down to 12 volts and use lighter-weight fixtures that can be easily moved.

If you're thinking of adding outdoor lights to an existing circuit, you'll need to make sure the circuit can handle the load. As a rule, a 15-amp, 120-volt circuit can handle a maximum of 1440 watts. Add up the watts marked on the bulbs and appliances fed by the circuit you want to add to. The difference between this sum and 1440 is the total number of watts you can add.

Taking wiring outdoors

Techniques for extending wiring to the outside are much the same as those used for extending wiring indoors, with two major exceptions: an outdoor electrical box must be weatherproof (for both 12-volt and 120-volt systems), and the circuit must be protected with a ground fault circuit interrupter type of circuit breaker (abbreviated GFCI or GFI).

A GFCI cuts off the power in 1/40 of a second if current begins leaking anywhere in the circuit. This fast reaction greatly lessens the chance of severe electrical shock. Even though you may plan to connect your lighting system to an existing outdoor receptacle, seriously consider replacing the outlet with one that has a built-in GFCI.

If you need help to add a new cir-cuit or to connect with an existing one, consult an electrician or refer to the *Sunset* books *Basic Home Wiring Illustrated* or *Home Lighting*.

12 volts or 120 volts?

Once you've brought your 120-volt indoor wiring to a waterproof housing box outdoors, you'll need to decide whether to add an outlet for a 12-volt transformer or to connect to 120-volt outdoor wiring.

12-volt system. Here are some reasons why you may wish to install a 12-volt system:

• Installation is simple: cable can lie on top of the ground, perhaps hidden by foliage; most fixtures connect to cables without need for stripping the wire insulation and splicing the wires, and no ground connections are required.

• No electrical permit is required for installing a system that extends from a low-voltage plug-in transformer (the most common kind).

• With low-voltage fixtures or wiring, there is much less danger that people or pets will suffer a harmful shock.

120-volt system. A 120-volt outdoor lighting system has these advantages:

• The buried cable and metallic fixtures give the installation a look of permanence.

• Light from a single fixture can illuminate a large area—especially useful for security and for lighting trees from the ground.

• Not only light fixtures, but also power tools, patio heaters, and electric garden tools can be plugged into 120-volt outdoor outlets.

Adding a 12-volt system

To install a 12-volt system of adequate capacity, you'll need a transformer, up to four 100-foot runs of two-wire outdoor cable, and a set of 12-volt fixtures. To activate the system, you'll need to connect the transformer, and perhaps a separate switch, to a 120-volt power source.

Wire size. Most low-voltage outdoor fixtures use flexible stranded-wire cable. The size of the wires in the cable will depend on the total wattage rating of the fixtures you plan to connect to the cable. Check the manufacturer's instructions for the correct size to use.

Types of transformers. Transformers designed for low-voltage outdoor lighting are mounted outdoors in weatherproof boxes; they have three-pronged (grounded) plugs that are inserted into outdoor grounded outlets or shockproof outlets protected by GFCI circuit breakers.

Most transformers are rated for home use from 100 to 300 watts. The rating shows the total allowable wattage of the fixtures served. The higher the rating, the more lengths of 100-foot cable—and consequently the more light fixtures—can be connected to the transformer. The maximum number of these cables is four.

Selecting a transformer that has a built-in timer or ON-OFF switch (some models have both) will relieve you from having to wire in a separate switch.

Installing a transformer. Most transformers for outdoor lights are encased in watertight boxes; to be safe, though, plan to install yours at least a foot off the ground in a sheltered location.

If you don't already have an outlet into which to plug the transformer, plan to install a GFCI-protected outlet. The drawing at near right, top, shows how to wire an outdoor GFCI outlet.

Many transformers have built-in switches—but some don't. Installing a separate switch indoors will probably prove more convenient than installing it outside. The drawing at near right, bottom, shows how to wire a new GFCI outlet to an indoor switch and existing power source.

To connect one or more low-voltage cables to the transformer, simply wrap the bare ends of the two wires in each cable clockwise around the terminal screws on the

Wiring a ground fault circuit interrupter (GFCI) outlet

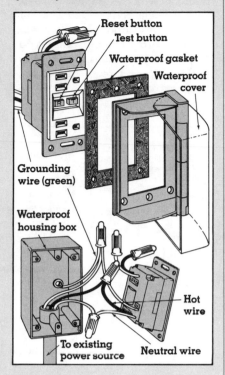

Reset button
Test button
Waterproof gasket
Waterproof cover
Grounding wire (green)
Waterproof housing box
Hot wire
Neutral wire
To existing power source

Wiring a new GFCI outlet to indoor switch & power source

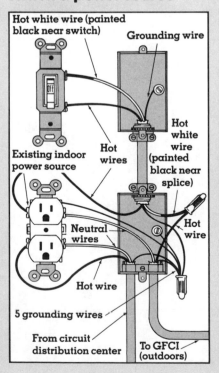

Hot white wire (painted black near switch)
Grounding wire
Existing indoor power source
Hot wires
Hot white wire (painted black near splice)
Hot wire
Neutral wires
Hot wire
5 grounding wires
From circuit distribution center
To GFCI (outdoors)

---

transformer (if the transformer accommodates more than one cable, the terminal screws will be arranged in pairs) and tighten the screws.

Connecting fixtures to the cable. Once your transformer is in place and you've decided where to put the fixtures, you'll need to hook them into the cable or cables leading from the transformer.

With some fixtures, you simply pierce the cable with a screw-type connector already attached to the rear of the fixture. With others, you must connect the short cable from the fixture to the main cable, using a screw-type connector. Neither of these types of connector requires removal of insulation from the cable.

A few brands of fixtures require splicing into the main cable. Use weatherproof plastic junction boxes to insulate splices that can't be pushed back into the fixtures.

Adding a 120-volt system

To install a 120-volt outdoor system, you'll need lighting fixtures, of course, and some underground 120-volt cable (if allowed by local code) or conduit—the length you can use depends on the size of the wire. You'll probably want to connect the system through an indoor switch and timer to an existing electrical source or circuit. Be sure to protect the circuit with a GFCI.

Installing an indoor switch & timer. By wiring in a switch and timer as shown in the drawing at right, you can turn lights on and off by hand or let the timer do it for you.

Types of outdoor cable & conduit. The size wire that you must use depends on the total wattage ratings of the lighting fixtures you'll connect to the system. Here are the maximum ratings for some typical wire sizes:

   #14 wire—1440 watts
      at 120 volts
   #12 wire—1920 watts
      at 120 volts
   #10 wire—2880 watts
      at 120 volts

Wiring an indoor switch & timer for 120-volt outdoor fixtures

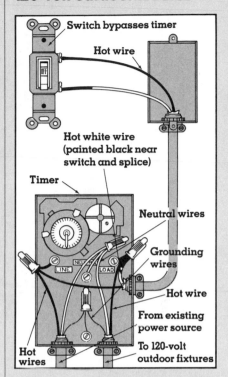

Switch bypasses timer
Hot wire
Hot white wire (painted black near switch and splice)
Timer
Neutral wires
Grounding wires
Hot wire
From existing power source
Hot wires
To 120-volt outdoor fixtures

Many local electrical codes require the use of rigid conduit for outdoor wiring. Plastic conduit, though lighter and less expensive than steel, must lie at least 18 inches underground. Steel conduit can be as little as 1 inch underground. Run two thermoplastic-insulated wires (TW) through steel conduit, which is self-grounding; run three TW wires (including a ground wire) through plastic.

Some local codes allow the use of flexible three-wire underground cable (UF) instead of rigid conduit. UF cable must be buried at least 18 inches deep. Work with UF cable in the same way you'd work with non-metallic sheathed cable. Before covering the cable with dirt, lay a redwood board on top of it so you won't accidentally spade through it at a later time.

For details about working with conduit and cable, as well as more information on installing outdoor electrical systems, see the *Sunset* book *Basic Home Wiring Illustrated*.

Garden work centers

The best garden work centers are planned to meet the specific needs of a home gardener, and usually they must fit into a limited space. The work centers on these pages were designed to suit a variety of situations. Adapt them to your needs, or let them inspire you as you design your own center.

Use decay and insect-resistant wood such as redwood, cedar, or pressure-treated lumber for any part of the work center that is close to or in contact with the ground. Make the work surfaces reasonably weatherproof, sturdy, and easy to clean. Though a gravel floor usually is adequate, don't overlook the possibility of brick or concrete—they're easier to keep clean.

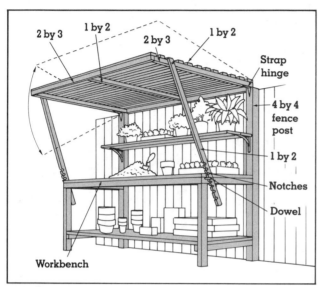

Attached to fence, workbench has hinged lath top that swings down to protect plants and can be adjusted to shade work area. Shelf below holds pots, seed flats, soil mix.

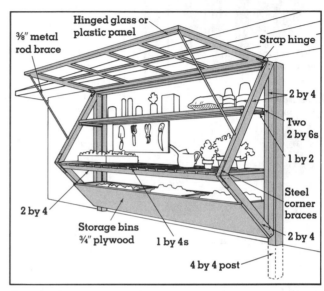

Attached to garage wall, workbench has hinged overhead with glass or plastic panels, plus storage bins beneath counter. Cover ends with plastic, lower overhead, and work center becomes a temporary cold frame.

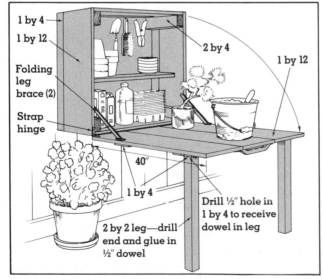

Wall-hung box with hinged front stores equipment and provides work table for potting and small jobs. Removable legs, held in place with dowels, are stored in cabinet when it's closed.

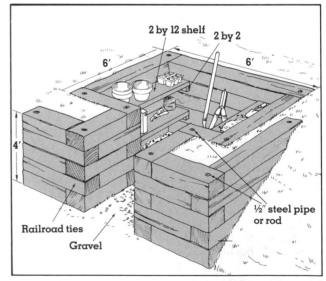

Railroad-tie bunker, set down in ground with planting beds mounded around it, provides hidden work center. Shelves of 2 by 12s supported on cleats provide work surface and storage area.

Concealed work center

This potting bench was designed to fit behind a 7-foot screening fence (see pages 64–68 for information on building fences), where it's shielded from the house and the display areas of the garden. You can adapt this bench to attach to an existing fence or to the side of the house or garage.

Make sure the posts are set in concrete deep in the ground (about 3 feet—or below the frost line in cold climates) to keep the roof (tilted at 11°) from toppling the fence. The roof shown is made of lath, but you could substitute rigid plastic sheet, corrugated plastic, or canvas (see page 79). Or you could use the lath and then cover it with one of these materials during the winter months. Such materials would not only keep the work surfaces dry but also would help protect plants against frost damage. If you prefer, you could cover just one section of the roof for protection against rain while working. Design: Osmundson-Staley.

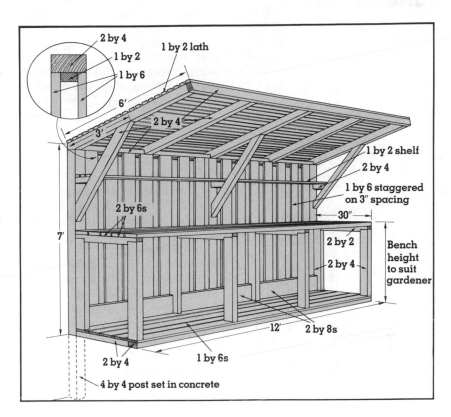

Work center in the round

Here's a work center that has it all—a translucent roof of corrugated plastic that lets in natural light, tempers the sun, and keeps out the rain; walls of inexpensive benderboard that help shade the interior and let in the cooling breezes; a workbench and a sink; and a smooth 10-foot-diameter concrete floor that's easy to keep clean. Design: William P. Bruder.

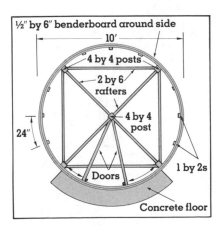

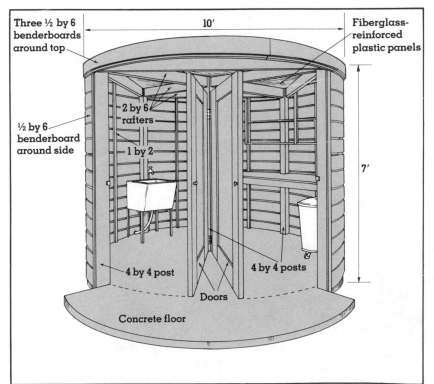

Garden storage

Question: How can you solve the problem of where to store lawn mowers, hoses, and other garden tools and supplies; cushions for outdoor furniture—and, during the winter months, the furniture as well; plus all the rest of the paraphernalia for outdoor living? Answer: Study the project ideas on these two pages, and adapt them to meet your specific needs. They include an A-frame storage shed with an adjoining arbor, a storage cupboard built against a fence or wall, suggestions for underseat storage, and some convenient hideaways for garden hoses.

A-frame storage shed & arbor

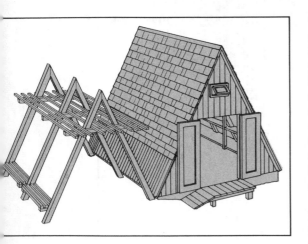

This combination storage shed and arbor—a pair of simple A-frame structures—gives a place to store all your garden tools, as well as such large equipment as a power mower, cultivator, and wheelbarrow. Shelves along one side hold pots and flats—or the top one can be used as a work counter. The shed even has a loft for storing the usual conglomeration of items no one ever seems to have enough space for.

Small windows on the front and back of the shed open for ventilation. The windows aren't the only sources of light, though—plastic panels in the sides admit diffused light. If you'd like more light—or less—alter the plan accordingly.

If storage isn't a need, you could use the structure, with its shady arbor, as a summer house for outdoor entertaining, as a playhouse for children, or even as a small weekend house. With little extra work and expense you could fasten canvas on the sides of the arbor to roll down, creating extra summer sleeping space. Design: Frank Shell and Dr. Rogers Smith.

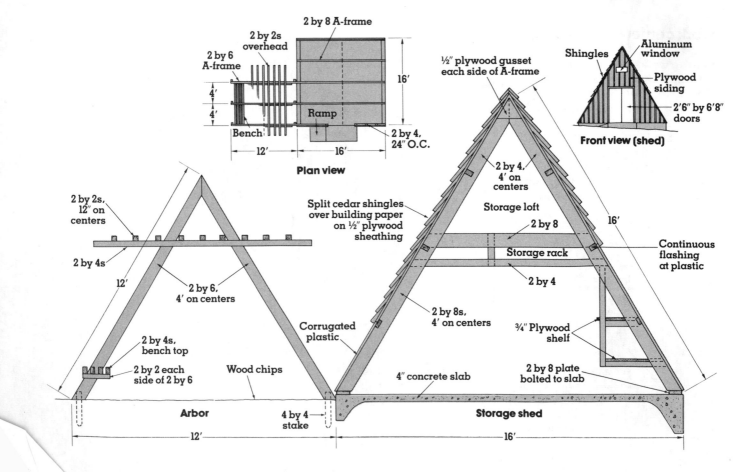

Gothic-style greenhouse

Bands of exterior plywood form the arches for this greenhouse, which is covered with thin plastic sheet to shed water and snow and resist wind. In late summer, you can convert the greenhouse to a propagating frame simply by replacing the plastic with shade cloth or lath. Attach either of the three coverings with staples or tacks.

Two small vent flaps—one at each end over the door—provide limited ventilation. For added ventilation on warm days, the doors can be opened or the greenhouse can be raised off the ground and supported on concrete blocks.

All lumber close to or in the ground should be redwood, cedar, or pressure-treated to resist insects and decay.

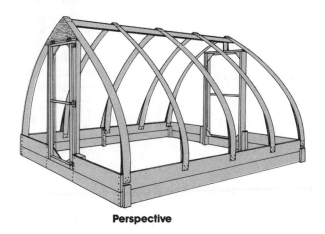

Perspective

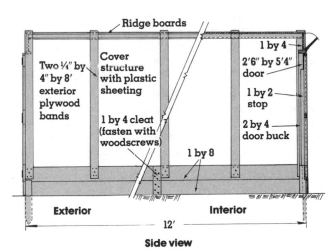

Side view

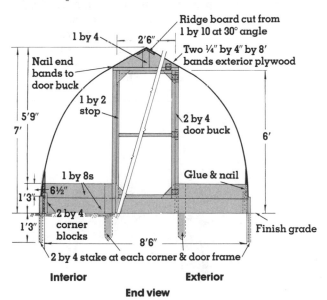

End view

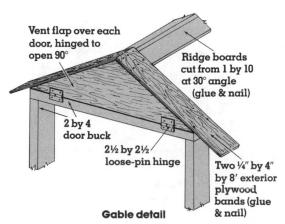

Gable detail

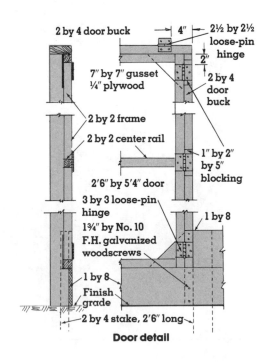

Door detail

Garden pools

Most people enjoy the sight and sound of water, and a garden pool can satisfy the desire to be near water, hear or see it, or even to dabble your feet in it. In addition, a garden pool can be a show place for aquatic and marsh plants, as well as a home for goldfish, koi, and other freshwater fish.

Pools—among the most challenging garden projects—can be built in any shape or size to suit anyone's taste, and from a variety of materials. One of the two pools illustrated here is built of brick, the other of reinforced concrete. A garden pool also can be built of concrete blocks or stone. Colorful ceramic tile can line a pool of concrete or concrete block. And don't overlook wood as a material for small pools.

Before you start building your pool, check with your building department for information on building codes and permits.

Two-section garden pool

Build this two-section pool in whatever size you wish, and grow water plants in one section, marsh or bog plants in the other. Though it's shown built of bricks, you can build it with concrete blocks (see pages 58–59)—just fill the cells of the blocks with mortar.

Excavate for the pool, making sure the bottom is level and the soil is well tamped. Do the plumbing for the drain, place the steel reinforcing, and then pour the concrete floor (see pages 36–37). Finish the concrete with a wood float and cover it with plastic to cure (see page 39).

Build up the walls with bricks, making sure they're well bonded (see pages 54–57); flush mortar joints are best for a pool. Build the exterior walls with a double row of bricks, and the wall between the pool and the bog with a single row. Near the bottom of the dividing wall, leave several weep holes by not mortaring some of the vertical joints. This will allow water to enter the bog and keep the soil wet. When you've finished the walls, build the coping around the outside edges with bricks laid flat and side-by-side.

After the mortar has cured, coat the inside of the pool with waterproofing compound. After it dries, fill the bog section with good, rich soil. Fill the pool with water, wait for the soil to become saturated, and you're ready to put in the plants. Design: Osmundson-Staley.

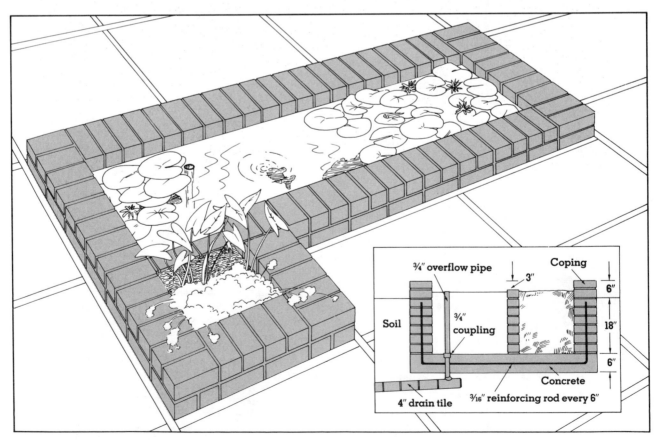

¾" overflow pipe Coping 3" 6"

Soil

¾" coupling

18"

6"

Concrete

4" drain tile ³⁄₁₆" reinforcing rod every 6"

Concrete pool

First, excavate and compact the pool site. If you plan to stock the pool with fish, allow for a finished depth of at least 18 inches to help discourage cats, raccoons, and other marauders. In areas of severe freezing weather, allow an additional 3 to 4 inches for a layer of gravel under the concrete. Slope the sides at 45 degrees. Allow for the perimeter lip that will support the stone coping.

Add reinforcing either by bending 6 by 6-inch wire mesh to fit inside the pool, or by using ¼-inch or ⅜-inch reinforcing rods. Bend the rods to follow the pool contours, and arrange them so that they resemble latitudinal and longitudinal lines. Space them 6 to 12 inches apart, tying the intersections securely with wire.

Using bits of brick or stone, support the reinforcing 2 inches off the earth or gravel. Then drive stakes in every square foot, with the tops extending at least 6 inches above the earth or gravel.

Mark the stakes 4 inches above the earth or gravel. Make your concrete by mixing 1 part cement, 2 parts sand, and 3 parts gravel with just enough water to wet all ingredients. Using a shovel or trowel, pack the mix firmly around the reinforcing up to the marks on the stakes. Remove the stakes and fill the holes with concrete. Finish the surface with a trowel; cure the shell (see page 39).

Finally, add a stone coping, mortaring the stones to each other and to the concrete lip. A small waterfall, as shown, adds a nice touch. A small, submersible, recirculating pump moves the water and provides a means of draining the pool. Route its power cord and supply tube up behind the waterfall. Connect the power cord to a properly installed outdoor outlet (see pages 102–103).

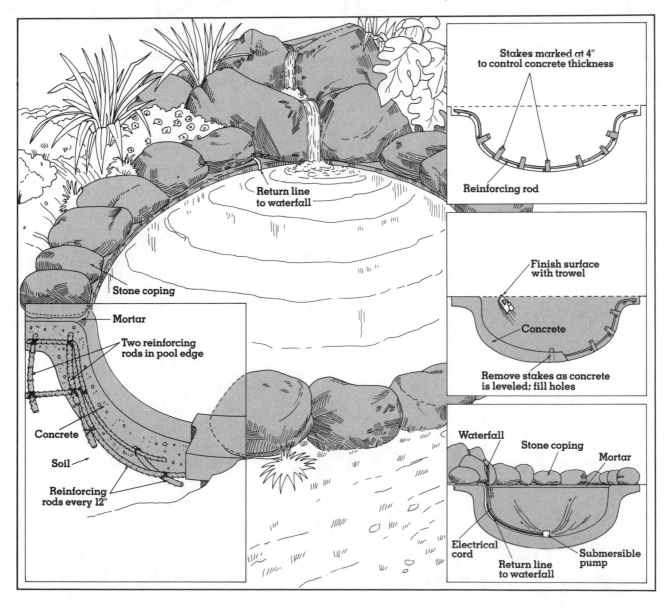

Return line to waterfall

Stone coping

Mortar

Two reinforcing rods in pool edge

Concrete

Soil

Reinforcing rods every 12"

Stakes marked at 4" to control concrete thickness

Reinforcing rod

Finish surface with trowel

Concrete

Remove stakes as concrete is leveled; fill holes

Waterfall

Stone coping

Mortar

Electrical cord

Return line to waterfall

Submersible pump

INDEX

Boldface numerals refer to color photographs.